Nanometrology Using the Transmission Electron Microscope

Nanometrology Using the Transmission Electron Microscope

Vlad Stolojan

Advanced Technology Institute, Department of Electrical and Electronic Engineering,
University of Surrey, Guildford, UK

Morgan & Claypool Publishers

Rights & Permissions
To obtain permission to re-use copyrighted material from Morgan & Claypool Publishers, please contact info@morganclaypool.com.

ISBN   978-1-6817-4120-8 (ebook)
ISBN   978-1-6817-4056-0 (print)
ISBN   978-1-6817-4248-9 (mobi)

DOI   10.1088/978-1-6817-4120-8

Version: 20150901

IOP Concise Physics
ISSN 2053-2571 (online)
ISSN 2054-7307 (print)

A Morgan & Claypool publication as part of IOP Concise Physics
Published by Morgan & Claypool Publishers, 40 Oak Drive, San Rafael, CA, 94903, USA

IOP Publishing, Temple Circus, Temple Way, Bristol BS1 6HG, UK

To Louisa, my beautiful wife.

Contents

Preface

Transmission electron microscopy (TEM) is synonymous with nanotechnology. It is the ultimate tool to see and measure structures on the nanoscale and to probe their elemental composition and electronic structure with sub-nanometre spatial resolution. The advent of spherical and chromatic aberration correctors has made TEM an incredibly powerful tool, combining spatial resolutions of 0.5 Å and energy resolutions of 10 meV, an energy resolution comparable to that of synchrotrons. This recent technological breakthrough will revolutionise not only our understanding of materials, but it promises to become a significant tool in understanding biological systems and bio-molecular systems. Viruses and DNA molecules, to name just two of nature's most significant nanoscale technologies, can now be studied with unprecedented spatial and energy resolutions, at the same time.

This book is aimed at the scientist looking to learn to use TEM as a tool to answer questions about physical and chemical phenomena on the nanoscale. It will help in understanding the type of information available and the requirements placed on the sample; it advises on how to plan the sample preparation and the experiments to obtain the best quality information and with what uncertainty. It also covers the majority of practical aspects in acquiring and processing images and spectra (x-ray and energy-loss).

This book grew out of a long experience of working on a wide variety of scientific problems, helping scientists and students alike to define the nanotechnology 'question' to ask of the TEM, the specific method to prepare the sample and which combination of imaging and spectroscopic techniques to use to answer that question, whilst ruling out artefacts. Seeing on the nanoscale is a voyage of discovery, one which could start right here.

Acknowledgements

I would like to thank D Cox, M Langridge, M Constantinou and G Rigas for their help and suggestions in preparing this manuscript. As with any such endeavour, I owe my knowledge and experience to a large number of excellent scientists and microscopists: LM Brown, A Bleloch, B Rafferty, D McMullan, JJ Rickard, N Menon, A Lupini, P Moreau, C Walsh, J Yuan, J Rodenburg, P Nellist, A Howie, A Ferrari, S Rodil, C Ducati, P Midgley, M Weyland, R Dunin-Borkowski, T Walther, J Sloan, S R P Silva and M Goringe. This is by no means a complete list, and I apologise to those whose name I have not included.

Author biography

Vlad Stolojan

Vlad Stolojan was born in Bucharest, Romania, where he spent the first 20 years of his life. He graduated with a BSc in Physics from the University of East Anglia, followed by a PhD in Physics from the University of Cambridge. His PhD work was on the Nanochemistry of Grain Boundaries in Iron, under the supervision of Professor L M Brown, at the Cavendish Laboratory, Cambridge, where he became an expert in Electron Energy Loss Spectroscopy in the Scanning Transmission Electron Microscope. After a short post-doctoral post in the group of Professor P A Midgley, at the Department of Materials Science and Metallurgy, University of Cambridge, he moved to his current place of employment, the University of Surrey. Currently he holds a Lectureship in Nanomaterials and Characterisation with the Advanced Technology Institute, University of Surrey. Further to electron microscopy techniques in nanotechnology, his research interests are in the growth and characterization of nanomaterials, particularly carbon nanotubes and graphene, as well as the focused ion-beam fabrication of micro-optical elements and electrospinning of nanomaterials into large-area, aligned-fibre sheets for electronic applications. He also provides day-to-day advice and training for electron microscopy applications, ranging from imaging of viruses in cancer cells to the analysis of ion-beam implanted layers in Si devices, hard-wearing coatings and Australian meteorites. He is an author of over 100 peer-reviewed articles based on electron microscopy and spectroscopy analysis and has filed for three patents on his electrospinning and micro-optical elements work. He is a member of the Institute of Physics and a fellow of the Royal Microscopical Society. He is a passionate skier, having been a ski instructor, and is a keen mountaineer and traveler. He is also a coach and regional-level player of korfball, a mixed-team sport similar to basketball, which he has played since university days (half-blue) and which led to him meeting his wife, Louisa.

IOP Concise Physics

Nanometrology Using the Transmission Electron Microscope

Vlad Stolojan

Chapter 1

Introduction

Seeing is believing, the old adage goes, but in microscopy of the nano-world, not everything that you see is believable. This is because a transmission electron microscope (TEM) operates in the world of quantum mechanics, where the electrons are described by a wavefunction, with amplitude and phase, which are changed by the sample and we *observe* its effect, which combines the amplitude and the phase. However, from explaining that the colour of medieval stained glass windows is due to plasmonic interactions at nanoparticles of gold and silver dispersed in the glass (Molera *et al* 2007), to imaging the growth process of carbon nanotubes *in situ* (Helveg *et al* 2004), TEMs have been instrumental in our understanding of nature and its processes on the nanoscale.

1.1 The TEM

A TEM uses electrons typically accelerated to energies of 60–300 keV to image, in transmission, a very thin sample. 300 keV electrons have a wavelength of ~4 pm, but the resolution of a TEM is ~1 Å due to spherical aberration, where rays entering the lens at different positions from the centre are focused in different places along the optic axis, rather than in one single focus point. The last two decades, however, have seen the introduction of spherical aberration correction, glasses for severely near-sighted microscopes, with significant order-of-magnitude improvements in resolution. The TEM uses electrons because they can be easily generated and manipulated using electric and/or magnetic fields, and they can be detected and counted; however, the TEM also requires a good vacuum (typically $<10^{-5}$ mbar), so that electrons do not scatter or get stopped but by interaction with the sample. The samples have to be very thin, with thicknesses of the order of 100 nm, so that electrons interact weakly with them.

Similarly to its visible-light counterpart, the TEM has probe-forming optics comprising (1) the electron source and a set of condenser lenses, (2) an objective lens, which is the main imaging lens, and finally (3) a set of projection lenses, which project an image of the sample or its reciprocal transform onto a screen, or a detector. With some designs of the objective lens being a twin-lens (one above, one below the sample), there are upwards of seven lenses within a basic TEM, with modern, aberration-corrected TEMs requiring independently controlled power supplies numbering in the hundreds, which can only be efficiently achieved via computer control and specialised software. For this reason, these instruments end up several times larger than their optical counterparts, requiring dedicated rooms and significant control of environmental parameters (temperature, air flow, vibrations, electric and magnetic fields).

When discussing imaging, we generally talk about separating features apart in position (the spatial resolution of the image), separating them in intensity (the contrast) and deciding whether features are real, artefacts or noise (signal-to-noise ratio). Once we identify the features, the challenge then is to understand what the observed two-dimensional (2D) features tell us about the three-dimensional (3D) sample. Deciphering the information requires an understanding of imaging as a result of the interaction of a fast, charged particle (the electron) with the material in the quantum mechanical sense (wave–particle duality). This means that the resulting image is a combination of scattering and interference, resulting in changes in the amplitude and the phase of the electromagnetic wave associated to the electron, as well as changes in its energy. What we record is the intensity, which maps the separate amplitude and phase information, a complex number, to a modulus-squared, real number, adding another challenge to the interpretation (one equation, two unknowns). Gabor (1948) solved this challenge through holography, a method that records both phase and amplitude information and it is perhaps significant that his Nobel prize was awarded in 1971, before that for the recognised designer of the first electron microscope (awarded in 1986), Ernest Ruska (Knoll and Ruska 1932). It follows that dark and bright features in a sample can be a result of scattering, for example because of different thicknesses, densities, or atomic numbers, i.e. a part of the material that is thicker, denser, or with higher Z will appear darker in the transmitted image, as it stops or scatters more of the transmitted electrons. However, dark and bright contrast also appears because of interference, where changing the optical paths through a sample can lead to constructive (bright) or destructive (dark) interference. So, although the current resolution of an electron microscope has reached below 50 pm (Erni *et al* 2009), the question of where the atoms (or more appropriately, columns of atoms) are in a crystal is not straightforward, without recovering both the amplitude and the phase. Figure 1.1 illustrates this point with high-resolution images of an AlAs tunnelling barrier in a GaAs matrix, where a change in the defocus of the microscope changes the interference conditions, reversing the contrast in the AlAs barrier region (see the areas marked by white squares in figure 1.1).

The answer to the question 'where are the atoms?' therefore is not straightforward, as the image recorded is a 2D projection of the 3D sample (the sample is integrated in the direction of travel of the electrons) and we record (observe) the intensity as the modulus-squared of the wavefunction, losing the amplitude and phase information. To determine the positions of a column of atoms or individual atoms, such as the quest for determining the positions and concentration of O atoms in YBCO high-T_c superconductors at grain boundaries (Houben *et al* 2006), we can use in-line holography, where we record a series of images at different foci and then iteratively reconstruct a 3D computer model of the structure. This technique, however, relies on 'guessing' a good solution for convergence and is more effective

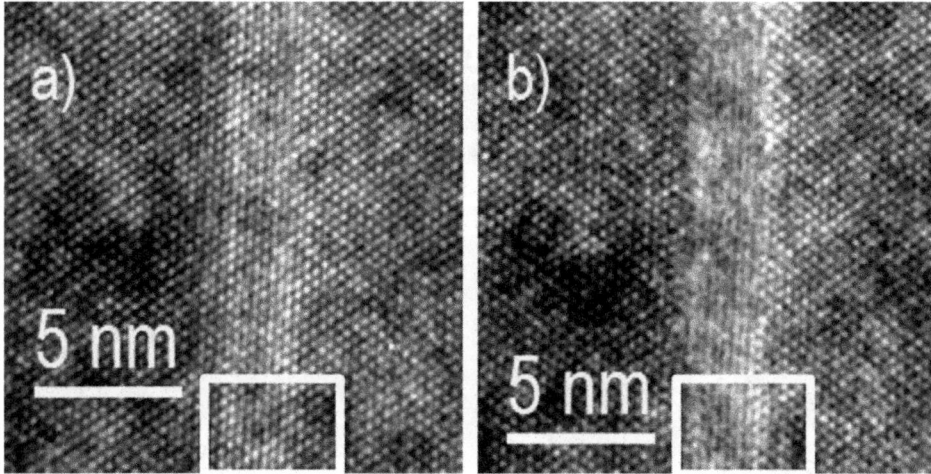

Figure 1.1. (*a*) The bright and dark features in the tunnel AlAs barrier in an asymmetric spacer layer tunnel (ASPAT) diode of GaAs reverse contrast when the interference conditions change in (*b*), illustrating that we need to know both the phase and the amplitude of the exiting wave to determine the positions of the columns of atoms.

for small, thin samples, such as materials encapsulated in single-walled carbon nanotubes, as seen in figure 1.2, where the arrangement of potassium and iodine atoms in the confined KI salt crystals can be determined.

For thicker samples, identifying single atoms in columns of thousands is only possible when the effect of that atom on the beam is significant (e.g. a strong scatterer) and the computing power required to simulate large volumes is significant. Another technique is based on manipulating the amplitude and phase information before recording the image such that, for example, the phase information is averaged in such a way that the image intensity then becomes proportional to the thickness, the density and the Z^2 (atomic number) of the sample; in this particular mode of imaging, we sum only over all high-angle diffracted beams with an annular detector. This mode of imaging is called high-angle annular dark-field (also known as HAADF or Z-contrast), a mode where the electron beam is focused to a spot on the sample and the image is formed by scanning the beam across the sample in a raster pattern, much as in a scanning electron microscope, an optical confocal microscope, or an atomic force microscope. In a regular TEM, the sample is imaged in parallel, rather than with a focused spot that is scanned. This scanning mode of imaging has allowed for the first identification of single atoms on a support, platinum on alumina (figure 1.3), by using the very high Z-contrast difference between them to balance the high concentration difference (Nellist and Pennycook 1996). In both methods, validation is sought through computer modelling and image simulation, with increased confidence brought in by aberration-correction and brighter, monochromated electron sources.

In some cases, we can use interference effects to our advantage, to enhance the contrast between layers of similar materials. For example, by changing the defocus, we can image amorphous carbon layers with different concentrations of sp^2-type bonds and hence different densities, but the same composition and thickness, as illustrated in figure 1.4.

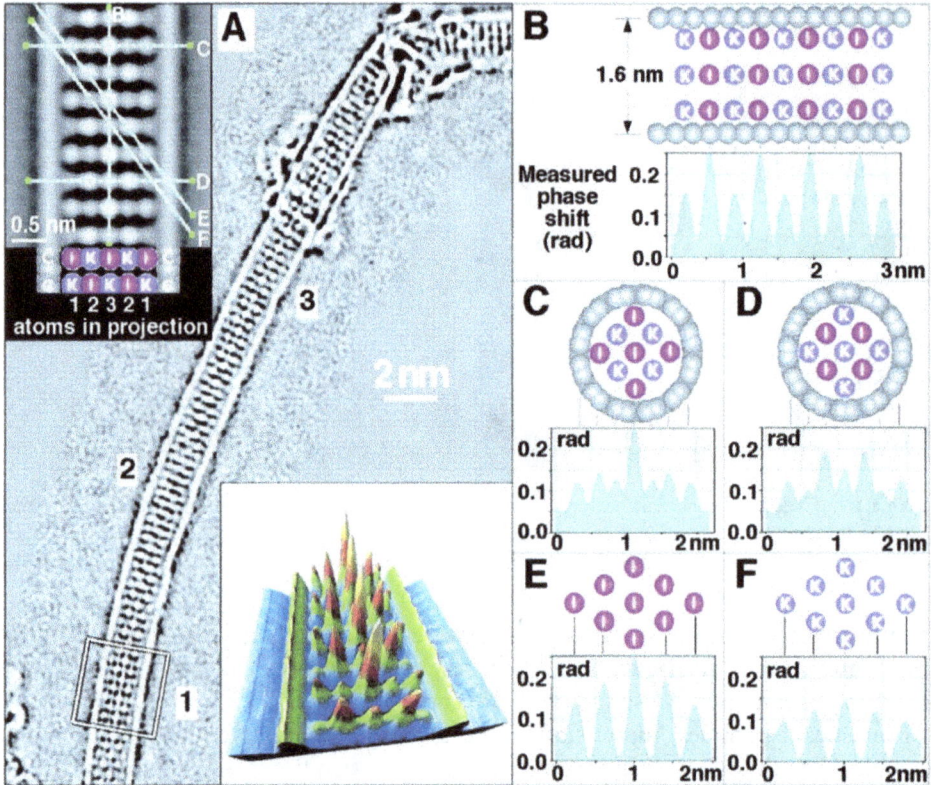

Figure 1.2. (*A*) Phase image showing a ⟨110⟩ projection of KI incorporated within a 1.6 nm diameter single-walled nanotube, reconstructed from a focal series of 20 images. Maximum and minimum spatial frequencies of 1/(0.23 nm) and 1/(1.05 nm), respectively, have been retained with a Wiener filter. The upper left inset shows an enlargement of region 1 (symmetrised about the chain axis) and a schematic illustration depicting the alternating arrangement of I–2K–3I–2K–I and K–2I–3K–2I–K 100 layers within the crystal. The lower right inset shows the surface plot of region 1. (*B*)–(*F*) Single-pixel line profiles obtained from line traces marked *B* to *F* in the upper left inset in (*A*). The background level in these profiles is arbitrary because the reconstruction procedure does not recover low-spatial-frequency variations in phase. Schematic crystal structures showing atoms contributing to the contrast are also shown. Image reproduced with kind permission from (Meyer *et al* 2000). Copyright 2000 the American Association for the Advancement of Science.

Again, in-line holography can be used to extract the mean inner potential across the device (Dunin-Borkowski 2000) and relate it to their resonant tunnelling behaviour and fast switching (Bhattacharyya *et al* 2006).

In everyday visible images, colour is the result of elastic (e.g. interference) or inelastic (e.g. absorption) scattering of photons from 3D objects. This colour then is processed and interpreted in our brains to identify objects and the materials they are made of, telling us, for example, if a vegetable is ripe. In the electron microscopy world, the in-joke is that 'all electrons are green', due to the original phosphor screen that used to convert their energy to visible wavelength photons. Following the interaction of electrons with samples, the electrons lose an amount of energy that can be measured using electron energy-loss spectroscopy (EELS). Furthermore, the energy transferred

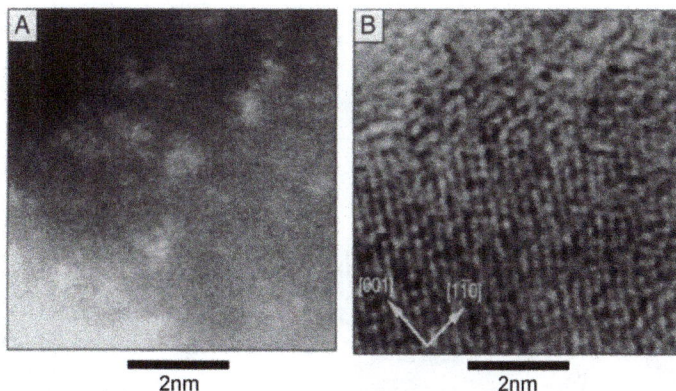

Figure 1.3. Simultaneously collected (*A*) *Z*-contrast and (*B*) bright-field images of a sample of Pt on γ-Al$_2$O$_3$. Some of the clusters in (*A*) can be seen to be resolved into single atoms. In (*B*), strong 222 γ-Al$_2$O$_3$ fringes can be seen with a spacing of 0.23 nm. Image reproduced with kind permission from (Nellist and Pennycook 1996). Copyright 1999 the American Association for the Advancement of Science.

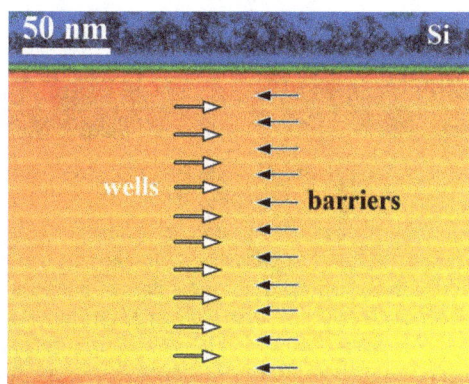

Figure 1.4. TEM image of an amorphous carbon superlattice grown on a Si substrate, with alternating layers of low sp^2-content amorphous carbon (wells) and high sp^2-content barriers visible when defocusing the microscope. The original greyscale image was recoloured using a rainbow look-up table.

to the sample can be further emitted as x-rays, Auger electrons, secondary electrons or light (cathodo-luminescence), as well as resulting in heating the sample, the induction of an electrical current, or even ejecting atoms from the sample (sputtering). All these (particularly x-rays and secondary electrons) contain information about the sample and can be routinely recorded at the same time as imaging, to reveal further information about it. As some of the energy absorbed from the incident electrons, or re-emitted as x-ray photons, results from atomic transitions (i.e. uniquely determined energy levels), we can identify the chemical composition and map its spatial distribution.

It is now possible to not only image with sub-atomic resolution, but also to record 'colour'—the energy lost by the electrons—in order to confirm the positions of columns of atoms and their chemical makeup, as seen in the point-by-point maps in figure 1.5 (Muller *et al* 2008).

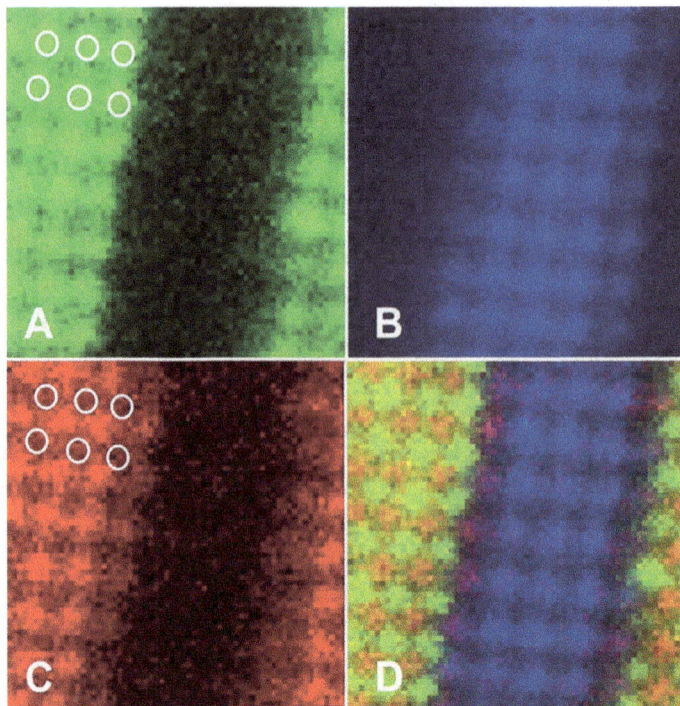

Figure 1.5. Spectroscopic imaging of a $La_{0.7}Sr_{0.3}MnO_3/SrTiO_3$ multilayer, showing the different chemical sublattices in a 64×64 pixel spectrum image extracted from 650 eV wide EELS spectra recorded at each pixel. (*A*) La M edge; (*B*) Ti L edge; (*C*) Mn L edge; (*D*) red–green–blue false-colour image obtained by combining the rescaled Mn, La and Ti images. Each of the primary colour maps is rescaled to include all data points within two standard deviations of the image mean. Note the lines of purple at the interface in (*D*), which indicate Mn–Ti intermixing on the B-site sublattice. The white circles indicate the position of the La columns, showing that the Mn lattice is offset. The live acquisition time for the 64×64 spectrum image was ~30 s and the field of view as 3.1 nm. Image reproduced with kind permission from (Muller *et al* 2008). Copyright 2008 1999 the American Association for the Advancement of Science.

Sub-atomic resolution is now possible due to the introduction of aberration-correction solutions by Rose (Haider *et al* 1998) and Krivanek (Batson *et al* 2002). The Rose corrector (CEOS) is now found on most large manufacturers' microscopes and was originally designed for TEM regular parallel imaging, whilst the Krivanek corrector was designed with the scanning imaging mode in mind and it led to the development of the Nion range of microscopes, currently boasting the highest spatial imaging and spectroscopy resolutions, down to being able to access vibrational spectroscopy information, normally reserved to optical spectroscopies (Krivanek *et al* 2013). Most electron microscopes are designed with magnetic lenses that produce a gradient magnetic field in the direction of travel of the electrons, which results in the electrons taking a cyclotron (spiral) path through the sample, to the detector. These lenses are incredibly poor; in optical terms, they are similar to using glass bottles to magnify features and a significant challenge to the microscopist is to minimise distortion through careful alignment of the lenses, the use of apertures

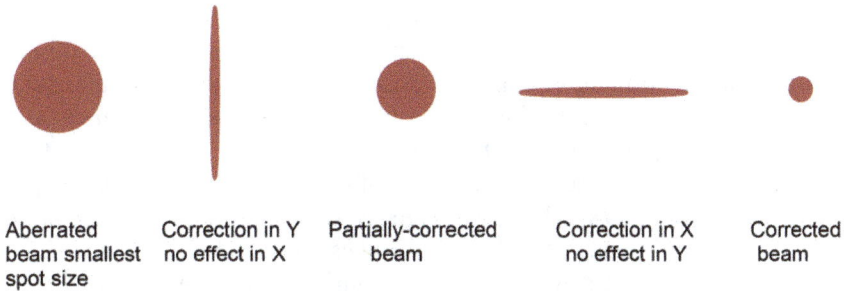

| Aberrated beam smallest spot size | Correction in Y no effect in X | Partially-corrected beam | Correction in X no effect in Y | Corrected beam |

Figure 1.6. Diagram of the principle of aberration correction, where a negative aberration can be obtained in one direction, at the cost of worsening aberration in the orthogonal direction. Shaping the beam to approximate a delta function minimises worsening of the aberration in that direction.

and the more recent aberration correction solutions. It is worth noting here that the spiral path of the electrons is useful in TEM alignment, as changes in lens strengths lead to image rotation, allowing for easier definition of lens optical centres and for the alignment of apertures with these centres. Unfortunately, unlike their optical counterparts, round magnetic lenses cannot be designed with negative spherical aberration; whilst optical microscopes were corrected in the late 18th century, the concept of aberration correction for electron microscopes had to wait nearly six decades from its inception for computer power to control all of the components of the aberration correctors. The TEM aberration corrector works by breaking the symmetry of the round lens through quadrupoles, hexapoles and octupoles, and using the fact that negative spherical aberration can be introduced in one direction, but at the cost of worsening it in the orthogonal direction. If the beam is, however, shaped so that it is very narrow (a delta function) in the worsening-aberration direction, then this effect is minimised; by now making it narrow in the other direction and correcting it again, by the time we reshape the beam we have an aberration-corrected beam (see figure 1.6).

The limits of spatial resolution are now in the realms of being able to place the electron probe within the exciton radius (Brown 1997) and with newly designed power supplies, which provide hundreds of kV, stable to meV, accessing vibrational spectra with high sampling resolution[1] is within reach.

We have already alluded to the fact that spatial resolution in electron microscopy is not a straightforward concept, as it is not just a property of the microscope, but also of the way the electron itself interacts and propagates through the sample, and we will revisit this in the following chapter in more detail.

Interaction of an electron with the sample is significant, ranging from radiolysis to sputtering. Electrons carry more than sufficient energy to overcome even the strongest atomic bonds, which can result in material being ejected, defects being

---

[1] A distinction is made here between sampling resolution, related to the size of the electron probe used, and the spatial extent of the specific electron–sample interaction. For example, plasmons are delocalised events, over the order of nanometers.

introduced or healed, can lead to atom diffusion and rearrangement, phase trans-formations, etc. Accounting for such damage is an essential part of electron microscopy analysis and it is common practice to first look to rule out beam damage before moving on to interpret observed changes or features as being representative of phenomena in the sample.

One recent interesting aspect in the design of modern microscopes is the push for lower accelerating voltages, which is driven by research into the new class of 2D materials, such as graphene, as well as nanotubes and nanowires, where the highly energetic beam can induce significant damage during imaging and analysis. Even in traditional hard material samples, significant damage to the sample can occur, particularly during spectroscopies where the beam dwells in a spot for long times, of the order of seconds (EELS) or tens of seconds (energy-dispersive x-ray spectroscopy (EDX)). Aberration-correction again comes to the rescue, allowing for atomic resolution at 60 keV (Pan *et al* 2014) and for the detection of specific spectroscopic signals within exposure times of the order of hundreds of miliseconds, as opposed to tens of seconds; this is, of course, backed up by the development in the spectroscopy detectors, such as direct detection of electrons and solid-state x-ray detectors.

All samples experience damage under the electron beam to a certain extent. In beam-sensitive samples, including biological samples, further mitigating measures need to be taken, such as cooling the sample to cryo temperatures, or limiting the dose either by blanking the beam or by scanning it rapidly to form an image. The sample needs to be dried carefully, so that moisture and carbon-containing organic residue (e.g. from solvents, such as acetone, methanol and IPA) desorb from the surface of the sample; moisture and residue tend to decompose under the beam and lead to chemical reactions that result in the build-up of amorphous carbon under the beam. Most microscopes have a cold-finger, a Cu plate sitting just above the sample and cooled to liquid-nitrogen temperatures, that helps to trap the surface-adsorbed species and reduce the amount of amorphous carbon build-up under the beam. Removal of the build-up is also possible using gentle plasma treatment or UV ozone cleaning, provided that the sample can withstand the process.

Electron-beam-induced damage is not always detrimental, with amorphous carbon build-up being able to act as a lithographic mask (Egerton *et al* 2004), whilst the dislocation and rearrangement of atoms can be used to weld structures, such as carbon nanotubes, induce surface ad-atom migration (Banhart 1999, Rodriguez-Manzo *et al* 2012, Cox *et al* 2005) or even study the catalytic growth process of carbon nanostructures in reverse (Stolojan *et al* 2006).

Ultimately, sample preparation is arguably the most important part of electron microscopy, with an often-quoted postulate that a microscopist spends 80%–90% of their time preparing the sample and only 10%–20% analysing it in the electron microscope. This is certainly true where a bulk sample needs to be reduced to a thin disk with a thickness of around 100 nm and as low as 10–20 nm for very high atomic resolution. 'Thinner is better' for high-resolution imaging, but the spectroscopic signal (EELS, EDX) requires a thickness that is usually around 0.5–1 times the mean-free-path (mfp) for inelastic scattering. Understanding the artefacts that can be intro-duced within sample processing, such as preferential grain boundary etching or strain-induced segregation of a constituent chemical to the surface of the sample, as well as taking into account electron-beam damage are an essential part of electron microscopy. Usually, one rules out all possible beam-induced defects and sample preparation artefacts before even considering that an observed feature is related to an original treatment or process, or a real feature of the sample. This skill is usually acquired only through experience and it is common to present one or a few null-tests for comparison (so-called 'sanity-tests'), to support the conclusions of electron microscopy analysis and interpretation.

For nanomaterials that are in a bulk agglomerate or in a solution, the sample preparation is relatively straightforward, where a holey-carbon grid is generally used to filter the nanomaterials out of the solution, either directly, or with the agglomerate first being dispersed in a solution. Nanomaterials generally end up suspended over holes and can be imaged without having to consider the influence of the amorphous carbon substrate. Imaging a sample through the substrate however (e.g. carbon nanotubes) can be beneficial in reducing the damage, as it provides heat-transport and electron-grounding pathways for the material of interest. The biggest challenges with this method of sample preparation though are the solvent residue as well the possible presence of surfactants, commonly used to keep nanoparticles from aggregating (e.g. quantum dots and metal nanoparticles) or to make nanomaterials hydrophilic (e.g. carbon nanotubes, graphene). These tend to form a small, mostly amorphous coating on the surface of the nanomaterials and can result in degradation and contamination under the beam. As this builds up into an amorphous layer, it produces a random change in the phase of the beam which blurs the information that comes from the crystalline part of the nanomaterial within.

For bulk crystal materials that need to be sliced thin, for example for imaging quantum layers buried in a semiconductor in cross-section, there are a number of mechanical and ion-polishing methods designed to remove material in a controlled manner. Ultimately, they all introduce a number of defects and artefacts, and their success rate is relatively low, depending on the durability (hardness, internal strain, etc) of the material and the skill of the preparator. The various techniques and approaches to produce a thin lamella and reduce the amount of damage introduced by the preparation technique are covered in numerous publications and here we will very briefly describe some of the major techniques. Mechanical cutting (to get the sample to a size that fits in the microscope holder), grinding and polishing are widely used for creating cross-sectional or plan-view samples, where a piece of the sample is progressively thinned by grinding with increasingly finer paper and then polished down to the best possible finish using a paste of very small diamond particles. If one uses a tripod polisher, which controls accurately the orientation of the polishing plane (in geometry, a plane is uniquely defined by three points, or one line and one point that does not sit on the line), the sample can be thinned directly down to a few tens of nanometres. Most common though, due to the fragility of samples and the significant internal strains when very thin, is to grind and polish the sample down to a few tens of microns and then use an angled ion beam (4°–5°) to further polish the central region of the sample until a small hole appears in the middle of it. The electron-beam thin region will be in the very near vicinity of the hole and its size will be dictated by the angle of the beam. A shallower beam angle reduces the amount of material removed significantly and is generally used for short periods to remove the surface layer of the sample that is damaged by ion-bombardment at the slightly steeper angles. The entire process is quite long (including the time for mounting glues to set), requiring patience and attention throughout, as well as a significant amount of dexterity when mounting, dismounting and handling the sample.

The advent of the focused ion beam microscope (FIB) has brought in a new versatility in sample preparation, by being able to target specific areas from which a thin foil can be lifted, by sculpting it out of the bulk material with energetic gallium ions. Several lift-out methodologies have been devised for thinning, removing the foil from the bulk material and attaching it to the TEM grid support (Eberg *et al* 2008). During FIB sample preparation, a relatively thick skin (10–50 nm) of damaged, amorphised material results on both sides of the foil and these can be further minimised using a number of strategies such as reducing accelerating voltages as the sample becomes thinner and using a very brief polish in the FIB at a slight angle (3°–4°) from the normal cutting one (Suess *et al* 2011).

This chapter has provided a brief introduction to the TEM and the concept of the recorded image contrast as resulting from a combination of amplitude and phase. It has also introduced the aberration corrector and briefly discussed sample–beam interaction and damage, as well as sample preparation. Chapter 2 will look specifically at imaging techniques and image resolution and chapter 3 will describe the most common spectroscopic techniques used in a TEM, EDX and EELS.

References

Banhart F 1999 Irradiation effects in carbon nanostructures *Rep. Prog. Phys.* **62** 1181–221

Batson P E, Dellby N and Krivanek O L 2002 Sub-angstrom resolution using aberration corrected electron optics *Nature* **418** 617–20

Bhattacharyya S, Henley S J, Mendoza E, Gomez-Rojas L, Allam J and Silva S R P 2006 Resonant tunnelling and fast switching in amorphous-carbon quantum-well structures *Nat. Mater.* **5** 19–22

Brown L M 1997 A synchrotron in a microscope *Biennial Mtg of the Electron Microscopy and Analysis Group of the Institute of Physics (2–5 September 1997, Cambridge, UK)* (Bristol: IOP) pp 17–22

Cox D C, Forrest R D, Smith P R, Stolojan V and Silva S R P 2005 Study of the current stressing in nanomanipulated three-dimensional carbon nanotube structures *Appl. Phys. Lett.* **87** 033102

Dunin-Borkowski R E 2000 The development of Fresnel contrast analysis, and the interpretation of mean inner potential profiles at interfaces *Ultramicroscopy* **83** 193–216

Eberg E, Monsen A F, Tybell T, van Helvoort A T J and Holmestad R 2008 Comparison of TEM specimen preparation of perovskite thin films by tripod polishing and conventional ion milling *J. Electron Microsc.* **57** 175–9

Egerton R F, Li P and Malac M 2004 Radiation damage in the TEM and SEM *Micron* **35** 399–409

Erni R, Rossell M D, Kisielowski C and Dahmen U 2009 Atomic-resolution imaging with a sub-50-pm electron probe *Phys. Rev. Lett.* **102** 4

Gabor D 1948 A new microscopic principle *Nature* **161** 777–8

Haider M, Uhlemann S, Schwan E, Rose H, Kabius B and Urban K 1998 Electron microscopy image enhanced *Nature* **392** 768–9

Helveg S, Lopez-Cartes C, Sehested J, Hansen P L, Clausen B S, Rostrup-Nielsen J R, Abild-Pedersen F and Norskov J K 2004 Atomic-scale imaging of carbon nanofibre growth *Nature* **427** 426–9

Houben L, Thust A and Urban K 2006 Atomic-precision determination of the reconstruction of a 90 degrees tilt boundary in $YBa_2CU_3O_7$-delta by aberration corrected HRTEM *Ultramicroscopy* **106** 200–14

Knoll M and Ruska E 1932 The electron microscope *Z. Phys.* **78** 318–39

Krivanek O L, Lovejoy T C, Dellby N and Carpenter R W 2013 Monochromated STEM with a 30 meV-wide, atom-sized electron probe *Microscopy* **62** 3–21

Meyer R R, Sloan J, Dunin-Borkowski R E, Kirkland A I, Novotny M C, Bailey S R, Hutchison J L and Green M L H 2000 Discrete atom imaging of one-dimensional crystals formed within single-walled carbon nanotubes *Science* **289** 1324–6

Molera J, Bayes C, Roura P, Crespo D and Pradell T 2007 Key parameters in the production of medieval luster colors and shines *J. Am. Ceram. Soc.* **90** 2245–54

Muller D A, Kourkoutis L F, Murfitt M, Song J H, Hwang H Y, Silcox J, Dellby N and Krivanek O L 2008 Atomic-scale chemical imaging of composition and bonding by aberration-corrected microscopy *Science* **319** 1073–6

Nellist P D and Pennycook S J 1996 Direct imaging of the atomic configuration of ultradispersed catalysts *Science* **274** 413–5

Pan C T, Hinks J A, Ramasse Q M, Greaves G, Bangert U, Donnelly S E and Haigh S J 2014 In-situ observation and atomic resolution imaging of the ion irradiation induced amorphisation of graphene *Sci. Rep.* **4**

Rodriguez-Manzo J A, Krasheninnikov A V and Banhart F 2012 Engineering the atomic structure of carbon nanotubes by a focused electron beam: new morphologies at the sub-nanometer scale *Chem. Phys. Chem.* **13** 2596–600

Stolojan V, Tison Y, Chen G Y and Silva R 2006 Controlled growth-reversal of catalytic carbon nanotubes under electron-beam irradiation *Nano Lett.* **6** 1837–41

Suess M J, Mueller E and Wepf R 2011 Minimization of amorphous layer in Ar^+ ion milling for UHR-EM *Ultramicroscopy* **111** 1224–32

Chapter 2

Imaging

In this chapter, we will further discuss the modes of imaging in a TEM and the information that can be recorded. In the previous chapter, we have explained that the contrast of columns of atoms can change from bright to dark (figure 1.1), and vice versa, depending on the imaging conditions, such as the defocus of the objective lens. A high-resolution image in this context is an image where the contrast is dominated by changes to the phase of the electron beam, such as those caused by elastic scattering from lattice planes, rather than changes to the amplitude. To understand how this happens, and also how we define the concept of the spatial resolution of a microscope, we need to first describe how the electron wavefunction propagates through the specimen and the microscope, and how the contrast is transferred, both as a complex quantity, but also how it is recorded as a real quantity. We will do this by defining later in this chapter the contrast transfer function and the transfer function, respectively. However, the theory developed makes an important assumption about the sample: that it can be represented as a phase object, with amplitude set to unity. As is common in many other theories, we then assume that the phase change is 'weak', meaning that it is small enough for us to series-expand the exponential function that describes absorption and neglect higher-order terms. This approximation holds for very thin samples and it is a reason why, in practice, very high resolution aims for specimens of a few tens of nanometres. Structures such as carbon nanotubes are ideal for this type of imaging.

We have introduced in the first chapter the concept of a TEM image as recording the intensity of a complex quantity, with amplitude as the real component and the phase as the imaginary component. Here we expand on this concept by introducing the convergent lens as a Fourier transform calculator, where the Fourier transform of the object is obtained in the back focal plane of the convergent lens, and the inverse Fourier transform (i.e. the magnified object) is obtained further away, in the imaging plane. In other words, placing the screen[1] in the back focal plane records

---

[1] By 'screen' we refer to recording the intensity at that plane; it can be a real or virtual screen, where a virtual screen is what the projection system would image onto a real screen.

doi:10.1088/978-1-6817-4120-8ch2
2-1

the diffraction pattern, whilst moving the screen further back records a magnified image. In this context, in the back focal plane of the objective lens we have the object in reciprocal space and we can consequently operate on particular spatial frequencies to perform operations such as band-pass and spatial filtering. In particular, we can select various features in the diffraction pattern with an aperture resulting in filtered images, such as bright-field (the 0th order) or dark-field images (higher orders of diffraction). A bright-field image has mostly bright contrast, with dark areas where electrons have been strongly scattered out into directions outside the selecting aperture, such as by oriented crystal planes, whilst dark-field images show bright contrast only in the regions of the sample that scatter strongly into a specific direction, as selected by an aperture placed in reciprocal space. A similar spatial filtering can also be performed on an unfiltered TEM image, usually a high-resolution image, where a virtual aperture can be used on the computer-calculated Fourier transform of the image (figure 2.1).

To understand what information is contained in the contrast of an image, particularly when dealing with nanomaterials, we need to understand how the information is transferred and recorded, and what the resolution achievable in the image is. Before discussing the concept of the resolution of an image, we need to introduce a further concept that describes how the information is transferred from the sample by the wavefunction of the electron beam through the lenses and onto the detector. Each of the spatial frequency components u of the image, in the back focal plane of the objective lens, has a phase associated with it, which is described by the contrast transfer function, $CTF(u)$. $CTF(u)$ is determined by the size of the objective aperture[2] $A(u)$, the envelope function $E(u)$ that describes the attenuation of the wave due to the finite source size (spatial coherence) and chromatic aberration, and the aberration function, $\chi(u)$, governed by the spherical aberration C_s and the

Figure 2.1. (*a*) Image of a ceria nanoparticle and its associated diffractogram (*b*), showing several diffraction spots, indicating that fringes are present in (*a*), even if not easily distinguished by eye. Placing a circular mask over one of the reflections and performing the inverse Fourier transform in (*c*) reveals areas of bright intensity where these spots originate from. This indicates that the nanoparticle is an agglomeration of smaller, single crystal entities ~15 nm in size. The images were collected using a Hitachi HD2300A STEM and processed using Digital Micrograph™. Sample courtesy of Dr MA Baker, University of Surrey.

---

[2] For a circular aperture, $A(u)$ is described by the Airy function, which is very similar to a sinc(x) function.

defocus Δf of the objective lens of the microscope. For a thin sample, $CTF(u)$ is (Williams and Carter 2009):

$$CTF(u) = A(u)E(u)e^{i\chi(u)} \tag{2.1}$$

where, considering that the astigmatism has been corrected, we are looking at a contribution of defocus and spherical aberration to the aberration function:

$$\chi(u) = \pi\,\lambda\Delta f\,u^2 + \frac{1}{2}\pi\,C_s\lambda^3\,u^4. \tag{2.2}$$

Note that Δf can be negative so $\chi(u)$ can be minimised at a certain defocus (see Scherzer defocus, later in this section). CTF relates to the wave as a complex quantity, so we have to redefine the contrast function relevant to the recorded intensity from (2.1) as the intensity transfer function $T(u)$ (Williams and Carter 2009):

$$T(u) = A(u)E(u)\,2\sin\chi(u). \tag{2.3}$$

When $T(u)$ is negative, the atoms will appear dark against a bright background (i.e. positive phase contrast—positive where the beam goes through, negative where it is scattered out), whilst for a positive value of $T(u)$, the atoms will be bright against a dark background (i.e. negative phase contrast). This is because there is a further phase shift of $\pi/2$ due to diffraction.

With all these concepts introduced, we can now go on to define the resolution in an image. For example, we can look at what the highest spatial frequency we can record is, or what range of spatial frequencies is transmitted with the same sign for the phase.

2.1 Resolution

The original concept of resolution, the Rayleigh criterion, refers to how close two circular sources (e.g. two columns of atoms) can be so that they can be distinguished. It states the sources are resolvable if their reciprocal space representations (the Airy discs) are separated such that the maximum intensity of one coincides with the first minimum of the other. This sets the limit for visible wavelength microscopy with a single lens to (n is the refractive index and θ is the convergence semi-angle)[3]:

$$r = 0.61\frac{\lambda}{n\,\sin(\theta)}, \tag{2.4}$$

Whilst glass lenses have convergence semi-angles of $\sim70°$ (1.22 rad), electron microscopes have convergence semi-angles of the order of 0.1 mrad ($\ll1°$). For a 200 keV field-emission microscope, with a convergence of 0.11 mrad, this would give an ultimate resolution of about 0.14 Å, whilst opening up the convergence angle using an aberration corrector to 0.3 mrad could achieve 5 pm!

The concept of resolution can be somewhat confusing. As already mentioned, we can describe the microscope in terms of the 'bandwidth', the highest spatial

---

[3] The definition of diffraction-limited resolution varies between using the radius of the Airy disk (1.22λ) and the separation between two Airy disks so that the contrast between them drops to 0.7 of the peak intensity (0.61λ) or to 0.9 (0.5λ), which is just about discernible.

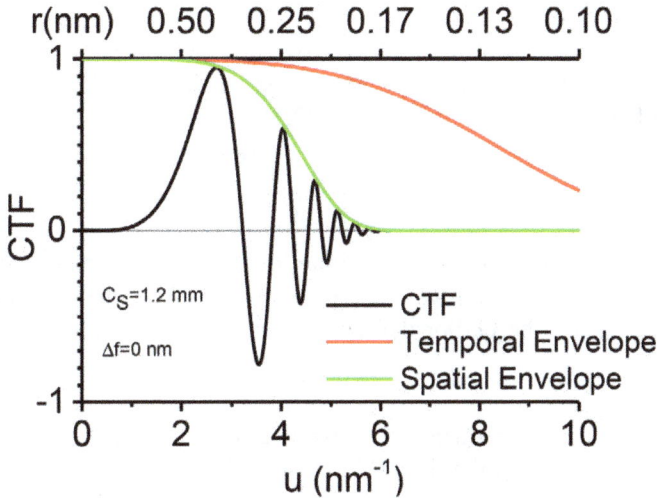

Figure 2.2. The contrast transfer function, calculated for a 200 keV field emission gun TEM at 0 defocus and 1 Mx magnification, oscillates as a function of the spatial frequency u. It crosses 0 the first time at 0.309 nm (point resolution), whilst the information transfer is limited by the spatial envelope to ∼0.166 nm (line resolution). Data calculated using ctfExplorer.

frequency that it can transmit. This is known as the *information limit*, or *line resolution*. This means that we can probably see detail in the recorded image up to this limit, but that does not mean the image is also interpretable. The concept of interpretability is directly related to the transfer function and the extent of spatial frequencies that are transferred with the same phase before the phase crosses through 0 and changes sign. This is the *point resolution* of the microscope.

Figure 2.2 shows the transfer function, calculated using ctfExplorer (Sidorov 2002), for a typical 200 keV field-emission microscope, without an aberration corrector, at zero defocus, also known as Gaussian focus. This is performed at 1 Mx magnification, which sets the reciprocal space width of the spatial coherence envelope (the higher the magnification, the lower the spatial frequency that the envelope extends to). This is counter-intuitive, as it means that when we increase the magnification, we decrease the ultimate resolution of the microscope. It implies that we do not want to increase the magnifying power of the microscope, but rather the sampling of the detector (i.e. the pixel count) if we want to push the microscope to its information limit.

Note also that, at Gaussian focus, large-scale features have no contrast, which is why, in practice, we find this condition by changing the focus of the microscope to seek the least contrast in the image. Defocus is measured then from this condition and most microscopes have a feature which allows the application of a pre-set defocus, to maximise the point resolution of the recorded image.

In-line holography uses the knowledge of the transfer function to extend the resolution of the microscope to the information limit.

The ideal form of the transfer function would be a step function, up to an information limit of u. Without going through the derivation, the closest we can get to the ideal curve (Williams and Carter 2009) is when the phase distortion function is close to $-120°$ and the function is flat (i.e. $d\chi/du = 0$). Scherzer (1949) defined this defocus as the optimum microscope resolution, a defocus that balances out the contribution of the spherical aberration in (2.3):

$$\Delta f_{\text{Scherzer}} = -\sqrt{\frac{4}{3}C_s\lambda}\,. \tag{2.5}$$

We can use the defocus to suit our imaging needs. For example, in off-axis holography we want to minimise the cross-talk between the amplitude and the phase images by using the Gabor defocus (Lehmann and Lichte 2002):

$$\Delta f_{\text{Gabor}} = 0.56\Delta f_{\text{Scherzer}}. \tag{2.6}$$

For maximum information throughput, we use the Lichte defocus (Lehmann and Lichte 2002), where u_{max} is the desirable information to be recorded:

$$\Delta f_{\text{Lichte}} = -0.75C_s\left(u_{\text{max}}\lambda\right)^2. \tag{2.7}$$

Figure 2.3 shows how the transfer function changes for the different defoci defined above, which suit specific imaging applications. For most high-resolution images, Scherzer defocus is optimum, whilst for some specific applications a defocus may be chosen to create passbands, optimised to image specific lattice fringes of a crystal. Furthermore, for a specific crystal, we can even define a defocus where gaps in the transfer function occur between Bragg reflections (Hashimoto *et al* 1977). This is fine for a perfect crystal, but if there is information where the transfer function crosses 0, such as would be related to boundaries and defects, then this will be lost.

Figure 2.4 shows a through focal series collected for the GaAs ASPAT diode also shown in figure 1.1, where the defocus values are measured from the 'least-contrast' image, as determined by the microscope operator and, as such, relatively subjective. Finding the 'least-contrast' image, or the focus point when using a microscope is equivalent to finding the peak of a relatively broad curve. The best measure of such a peak is the first derivative and the microscopist achieves this by the rate at which he/she changes the focus; the skill is to then stop this variation at the peak. The introduction of aberration correctors, which rely on diffractograms or ronchigrams[4] to measure the aberrations, has made this process less subjective. In figure 2.4, the image taken at Scherzer defocus (-45 nm for this specific TEM) is the one whose contrast is directly interpretable (although one must not forget the uncertainty in determining the Gaussian focus), whilst the 34 nm overfocus image transfers the most spatial information.

If we now examine the Scherzer-defocus image in more detail, in figure 2.5, we can extract information about the AlAs barrier, such as its width, 2.85 nm, but also the fact that it is not a flat, uniform layer, but appears with step defects (indicated by the arrow).

---

[4] The ronchigram is specific to probe-forming set-ups, such as the STEM, and is a directly projected image of the convergent beam through the sample onto a detector (also known as a shadow image). It contains information about the probe and the lens aberrations.

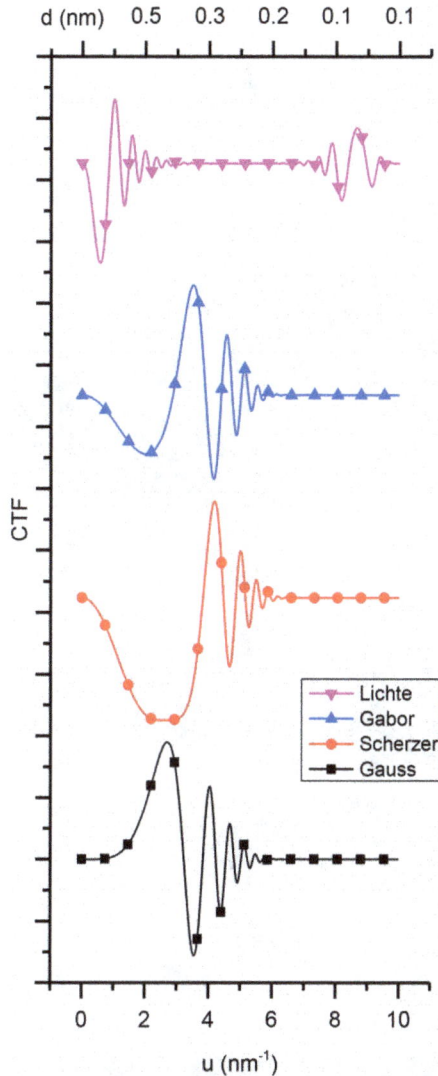

Figure 2.3. A comparison of the transfer functions for specific defoci that maximise the information transferred. Data calculated using ctfExplorer.

In chapter 1 we have discussed Z-contrast imaging as an incoherent image, where the phase information cancels out, an image being formed when the probe is focused and then rastered across the columns of atoms, whilst we record the high-angle scattered beams (usually, the inner half-angle of the annular detector is >80 mrad). This type of imaging allows for the direct identification of the columns of atoms, as seen in figure 2.6, where the atoms of GeTe crystals filling the inside of single-walled carbon nanotubes are visible in the Z-contrast image, seen here in comparison to the bright-field image, collected at the same time. The smallest probe size is related to the *point* resolution but, unfortunately, there is a further complication as the beam

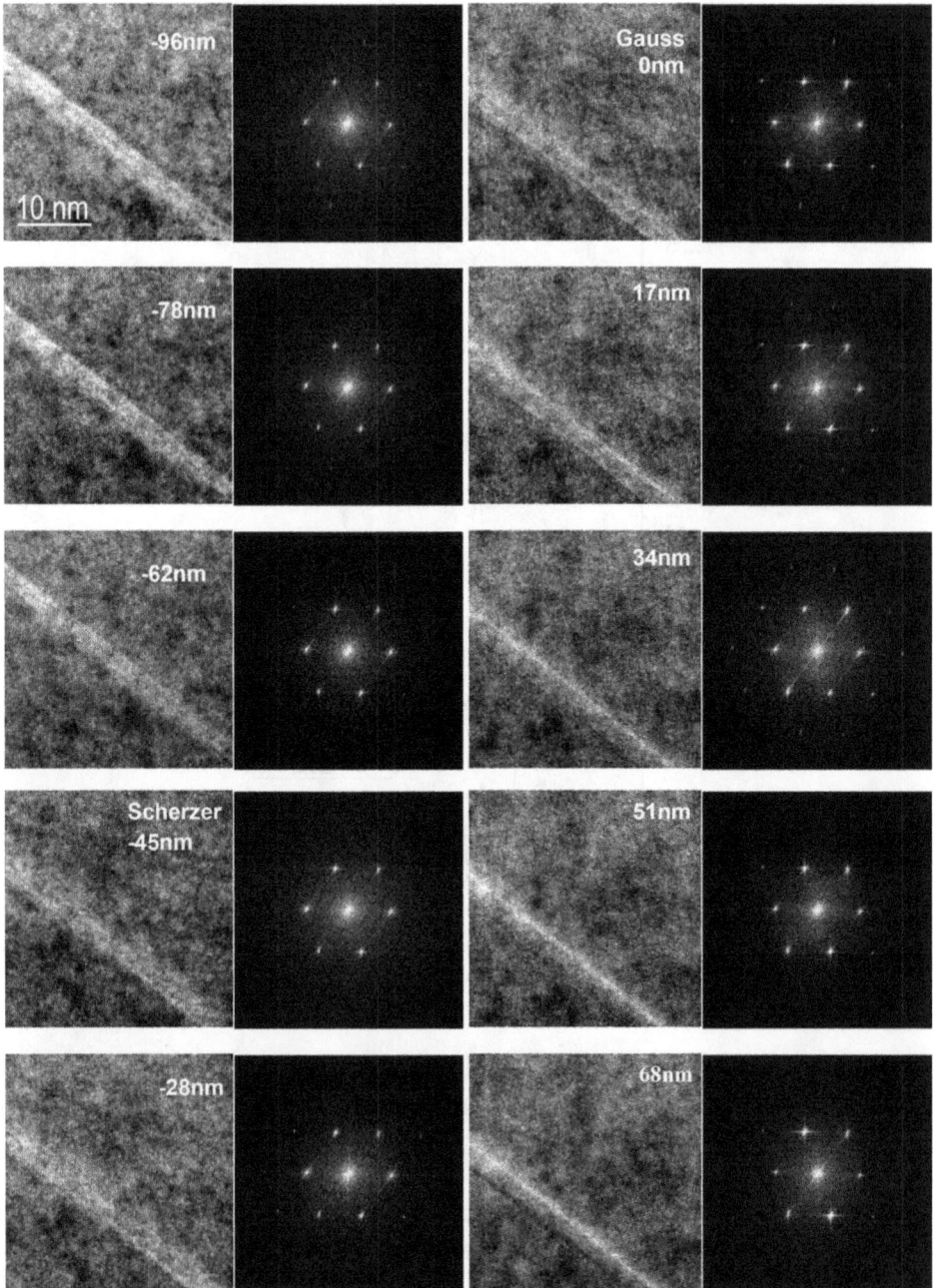

Figure 2.4. A through focal series of a GaAs asymmetric tunnel diode, showing the pair diffractogram to the right of the image. At Scherzer defocus, the image atomic contrast is interpretable, whilst higher spatial information is transferred at other defoci, as evidenced by the reflections in the diffractogram, e.g. 0 nm and 34 nm defocus. Images collected using a Philips CM200ST and processed using Digital Micrograph™.

Figure 2.5. The Scherzer-defocus image from the series in figure 2.4 can be used for layer thickness measurements to within an atomic layer and identification of defect regions (arrow). Image taken using a Philips CM200 TEM.

propagates through the sample. The electron probe 'channels' through the sample so that the intensity of the beam, even if very small (aberration-corrected) and placed between two columns of atoms, shifts within the first few layers of atoms to the neighbouring columns, spreading as it exits the specimen (Allen *et al* 2003).

In conclusion, in order to understand what can be measured and what can be interpreted in a high-resolution electron microscopy image, we need to have a good understanding of the various parameters of the microscope (aberrations, source size and stability, convergence angles) and use defocus as a tool to extract that knowledge, in combination with modelling and simulations.

2.2 Image calibration

In this section we discuss the repeatability and reliability of distance measurements in a TEM. The biggest challenge faced is that the lenses are magnetic and suffer from hysteresis. This means that, even if the lenses go through the hysteresis loop every

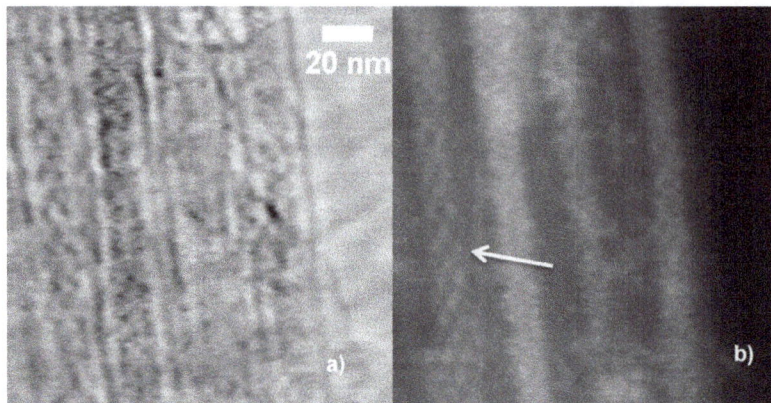

Figure 2.6. (*a*) Bright-field image of a bundle of single-walled nanotubes filled with GeTe. (*b*) The *Z*-contrast image shows the columns of the GeTe crystal inside the single-walled nanotube. GeTe is expected to be amorphous below 2 nm, but the nanotube confinement leads to crystallisation (Giusca *et al* 2013). Images obtained on a Nion-corrected VG501 STEM.

time their focus is changed, their magnetic fields can change slightly, modifying the magnification. Furthermore, going from overfocus to underfocus and back does not follow the same magnetisation path, which means that the direction from which we approach the chosen defocus (as described in section 2.1) is important. Usually the variations within an imaging session are small, but they are important when trying to measure crystal lattice fringes accurately and attempt to identify phases of the same compound. In practice, a 10% error is usually quoted on measurements, but this can be significantly improved if several lattice fringes can be measured, as there will be a linear correspondence between the measured fringe spacings and the tabulated values for the lattice spacings.

Because of the hysteresis of magnetic lenses, manufacturers have recently started moving towards focusing by moving the specimen, rather than changing the strength of the lenses, much as is done in conventional optical microscopy, supported by the very accurate motion technology of scanning probe microscopes, such as atomic force microscopes. This has improved the reproducibility and reliability of spatial measurements of TEMs. Nevertheless, having an internal calibration scale within an experimental session, such as a material with known lattice spacing, is always going to produce the most reliable results. Such materials can be single crystal substrates, such as Si, sputtered Au islands, etc. A word of warning though on using the spacing of the shells in carbon nanotubes, which can vary between 0.31 nm and 0.34 nm, depending on their quality (production method, treatments, etc).

The magnification scale is calibrated either by using lattice fringes in the direct image, or by using the corresponding reflections in the diffractogram (figure 2.7). The standard deviation of the calibration will scale down with the number of fringes counted. At the same time, it is worth checking the image for distortions and scale variations by imaging the same region of the sample, particularly at smaller magnifications. If the diffractogram is used for calibration, the standard deviation is determined by how accurately we can select the centres of the pair of reflections; it is recommended that a profile curve across the pair of reflections is taken and the peaks

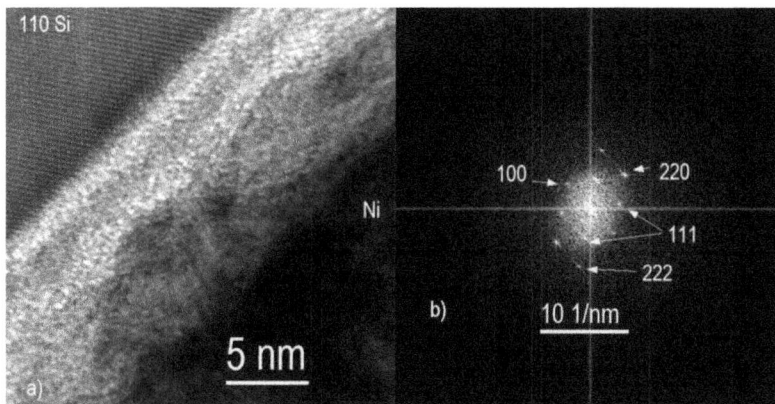

Figure 2.7. (*a*) A high-resolution electron microscopy image of a layer of Ni deposited on an amorphous carbon layer, on top of a 110 Si substrate. The Si substrate is used to align the sample so that the deposited layers are viewed parallel to the beam and is also used in calibration. The scale calibration can be performed by identifying lattice fringes and counting several of them. (*b*) The diffractogram of a section of image (*a*) containing the Si substrate is more useful at identifying the corresponding Si planes, as indicated by the labels on the figure. Calibration can be performed here as well. Note that 222 corresponds to 1.56 Å, below the interpretable resolution of this particular TEM, but within the information limit. The images collected using a Hitachi 2300A STEM and processed using Digital Micrograph™.

are fitted using Gaussian or Lorentzian functions. It is common to check the calibration of TEMs on a regular basis, at least monthly. The diffraction pattern is also very useful in determining the angular size of the apertures, where the angle associated with a reflection can be calculated directly from Bragg's law, which can then be compared to the size, recorded in the diffraction pattern, of the respective aperture. Knowledge of the angular size of the collector apertures at the entrance into the EEL spectrometer is, as we will see in chapter 3, essential for accurate, quantitative analysis.

The scanning TEM (STEM), the mode in which the Z-contrast mode introduced in chapter 1 is accessible does not suffer from the same hysteresis issues because the magnification is given by the area scanned by the scan generator and not the strength of the lenses. The scanning is achieved using electrostatic plates, where the applied voltages are highly accurate and reproducible. If the height of the sample is always brought near the optimum pre-sets for the microscope, then the calibration can remain reliable for much longer periods of time. As in the TEM case though, at lower magnifications image distortions can be introduced as the beam is scanned further away from the optical centre of the objective lens, where the effect of the aberrations becomes more pronounced.

If the images are recorded on third party cameras, or cameras on the end of an energy-loss spectrometer, the user must be aware of the communication between the microscope and various pieces of hardware, where the third party camera uses the microscope's value of the magnification to calculate the calibration scale, usually by interpolating between a couple of values at which the calibration was performed. This interpolation is assumed to be linear and can be improved by sampling it at more of the magnification pre-sets. Nevertheless, an internal calibration scale or a regular check are always advised.

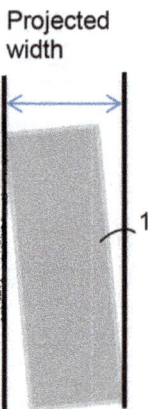

Figure 2.8. The projected width of a nominally flat layer misoriented by 1° depends on the thickness of the sample.

2.3 Sample orientation

With a reminder that the TEM image is a 2D projection of the 3D volume of the sample, performing accurate measurements of layers such as the quantum barrier layer in figure 2.5 requires accurate alignment of the said layer parallel to the beam. For example, the projected width of the 2.5 nm width of the quantum barrier layer for a 50 nm thick sample, misoriented by 1° as in figure 2.8 is: 50 sin(1°) + 2.85 cos(1°) = 3.72 nm, an increase of ~30% on the original thickness! Therefore, for accurate measurements, this orientation needs to be within ~0.1° of the layer–substrate interface, or the sample needs to be very thin (10 nm for a 6% error, for this specific case).

When focusing the probe, whether in a TEM or a dedicated STEM, the diffraction pattern of a single crystal shows some characteristic lines, called Kikuchi lines. They also occur in diffraction patterns of thicker specimens and are attributed to inelastic and elastic scattering of diffuse electrons by the thermal vibration of atoms in the single crystal. The thicker the crystal, the more intense these scattering events are and the more visible the lines are. These lines appear in pairs, one bright (excess) line and one dark (deficient) line, whose separation (in reciprocal space) is inversely proportional to $1/d_{hkl}$, where d_{hkl} is the lattice spacing of the hkl lattice planes (hkl are Miller indices). All the Kikuchi lines develop into a pattern (the Kikuchi pattern) of bands which intersect at major zone axes and offer a tilting route in a microscope to easily orient the sample onto a specific crystallographic direction, to within 0.1° (figure 2.9). Accurate tilting is also essential for high-resolution imaging, as well as other modes of imaging, where certain crystalline directions are chosen to diffract strongly. Kikuchi patterns are essential in generating orientation maps for polycrystalline samples using electron-backscattered diffraction (EBSD), a technique more commonly encountered with scanning electron microscopes.

A more complicated alignment process is necessary when looking at interfaces between crystals, such as grain boundaries. If the boundaries are planar over the thickness of the sample, they can be aligned parallel to the electron beam so that imaging and spectroscopic profile scans can measure changes occurring across the boundaries. It is common to study the segregation of impurities to the grain

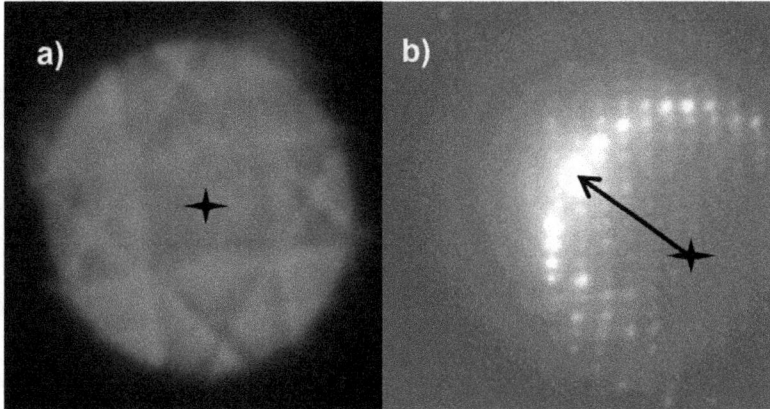

Figure 2.9. (*a*) Kikuchi bands cross at major zone axes and can be used for aligning a direction in the crystal (black cross) with the optic axis of the microscope to within 0.1°. (*b*) The diffraction pattern shows that the zone axis (black cross) needs to be tilted to bring it onto the optic axis, in the direction indicated by the arrow.

boundaries (Shibata *et al* 2004) and determining their spatial distribution and concentration is reliant upon accurate alignment with the optical axis.

For samples where no single crystal substrate can be used for tilt-alignment, it may be necessary to track the separation of two recognisable features as a function of tilt and find the optimum tilt from fitting the data. Alternatively, fiducial layers can be deposited and it is always useful to align the feature of interest parallel to the microscope's major tilt axis.

2.4 Modes of imaging and diffraction

We started this chapter by introducing the convergent lens as a Fourier transform calculator and discussing the contrast transfer function as defining resolution. The majority of the modes of imaging are spatial filtering techniques operated on the beam as it propagates through the sample, either in reciprocal space (the back focal plane of the objective lens), or on the image itself (e.g. by selecting an area, changing the focus).

2.4.1 *Z*-contrast imaging

In a TEM, the image is acquired in parallel, which means that one specific type of information (e.g. a bright-field image) is collected at a time; if the same area needs to be imaged under different conditions (e.g. a dark-field image), this needs to be acquired subsequently, either by using pre-sets to change the microscope's lenses and deflectors to a different setting, or by applying the changes manually. The sample, during this time, is continuously irradiated over a large area, so some damage assessment should be carried out on a sacrificial area of the sample. In STEM mode, either in a TEM or a dedicated STEM instrument, the image is collected serially, point-by-point, but several signals can be collected at the same time, such as secondary electrons, a bright-field image and a *Z*-contrast image, as seen in figure 2.6, as well as spectroscopy signals (EDX) and EELS (for which the detector is on the same axis as the bright-field image, so only one of them can be collected). Although it is true that each type of signal may

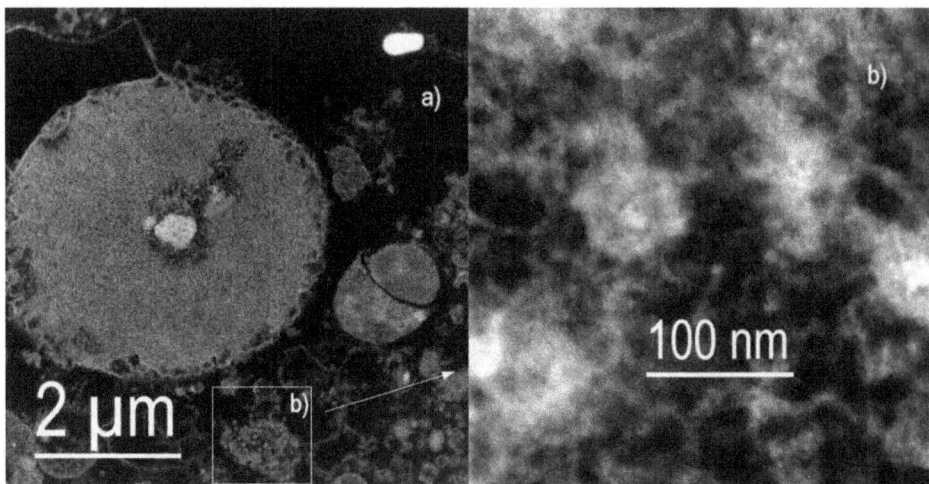

Figure 2.10. (*a*) Z-contrast image of a mouse intestine cancer cell, treated with a reo-virus, showing the nucleus. Note the very bright contrast of the particle at the top right, OsO_4, used to stain the sample, to enhance the contrast in TEM images. (*b*) Close-up Z-contrast image of the area marked in (*a*), showing the typical hexagonal head of a virus and DNA strands surrounding it. Images collected using a Hitachi HD2300A STEM. Sample courtesy of Professor H Pandha, University of Surrey.

require different exposure settings, they are available at the same time. The STEM mode concentrates a high current into a small spot but it dwells on the sample for a very short period of time, which may, in some cases, reduce the damage, such that biological samples can be imaged at 200 keV without cooling them to cryo temperatures.

Figure 2.10 is an example of the power of this type of imaging, demonstrating the ability to resolve down to DNA molecules (the white strands seen surrounding the typical hexagonal virus head), without any special measures to protect the sample from irradiation damage. It is also proof of the amazing work by Crewe (1964)—including the concept of imaging with energy-selected electrons to form colour images (see chapter 3)—which led to the development of the STEM (Crewe *et al* 1968) and showed its ability to image carbon-on-carbon, to distinguish DNA molecules supported on an amorphous carbon film (Crewe 1971).

When applied to multi-layered semiconductor devices, such as GaInAsP/InP quantum-well lasers in figure 2.11 (Adams *et al* 2015), Z-contrast can be used to characterise the molecular-beam epitaxy growth process and assess the compositional variations across the barriers and the wells, as well as any issues when switching from one mode of deposition to another, such as incomplete evacuation of a dopant species before the deposition of the next functional layer (indicated by the red arrows). These compositional variations on the nanoscale can lead to significant changes in the heights of the barriers and the depths of the wells, affecting the lasing properties of the device.

2.4.2 Simultaneous recording of multiple signals

Operating the electron microscope in STEM mode or using a dedicated STEM has a significant advantage in simultaneously using the multitude of signals available to

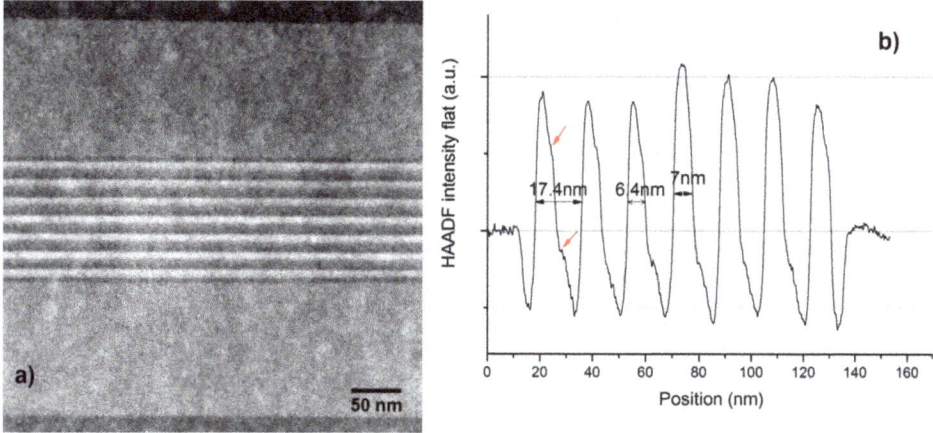

Figure 2.11. (*a*) *Z*-contrast image of an InP quantum-well laser, with quantum wells appearing as brighter lines due to the higher As content than the barriers (all other components are the same). (*b*) An integrated line profile across the structure in (*a*) shows the relative heights and depths of the barriers and the wells and the arrows show a less abrupt profile than expected from an immediate turn-off of the As supply. The sample has a slight thickness gradient, which has been removed from the profile. Sample courtesy of Professor A Adams and Dr I Marko, University of Surrey.

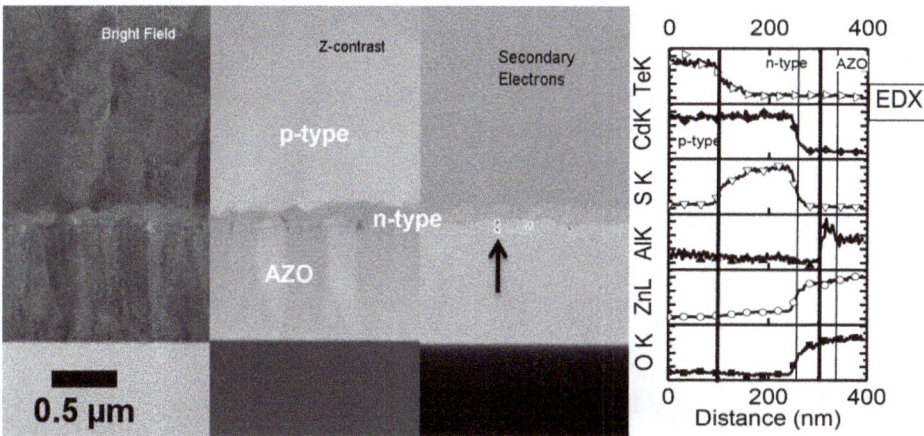

Figure 2.12. From left to right: bright-field, *Z*-contrast and secondary electron images of a CdTe solar cell grown on a transparent conductor, AZO. The secondary electron image shows that the thin layer next to the AZO has porosity, with only a couple of pores at the surface (indicated by the arrow in the secondary electron image). The bright-field image shows columnar growth of the AZO, whilst the *Z*-contrast image shows two compositionally distinct layers in the n-type region. When coupled with spectroscopy, such as EDX, the compositional profile in the fourth panel shows (from right to left) an aluminium-enriched surface to AZO, followed by ZnO, CdTeS, with S concentration decreasing towards the last CdTe layer (p-type). Sample courtesy of Dr M A Baker, University of Surrey.

understand the morphology of samples, such as the example shown in figure 2.12 of a CdTe solar cell. The sample here has been prepared using a FIB microscope, so it has a nearly uniform thickness throughout. This is advantageous when interpreting the *Z*-contrast, where the intensity is proportional to the thickness, as well as the density and Z^2. The bright-field image shows crystalline grains where the transparent conductor, aluminium-doped zinc oxide (AZO), revealing a typical columnar-growth structure. The secondary electron image shows that the surface of the sample is relatively smooth and flat, with a few pores in the ZnO layer just above the AZO. Comparing with the bright-field and Z-contrast images, which show more pores, we conclude that the layer itself is porous and this is not an artefact of ion-beam milling, such as grain pull-out. This also means that the contrast in the Z-contrast image can be interpreted as changes in Z^2 and density. Note that some crystalline contrast is still present, which means that the inner angle of the annular detector should be increased further, if the conditions (microscope, signal intensity) allow. Although we have not specifically introduced EDX, we have mentioned it in chapter 1 as one of the ways in which we can determine the 'colour' of electrons. In the case of this solar cell (figure 2.12), we can perform a point-by-point scan across the n-type region of the cell, and can determine further important features, such as the Al enrichment at the surface of the AZO and the two-step profile of the sulphur concentration, decreasing near the p-type region.

2.4.3 Bright-field imaging and the diffraction pattern

In a TEM, the imaging and diffraction modes of operation are accessible as pre-sets, where the projector systems switches between imaging the back-focal plane of the objective lens and the image plane.

Figure 2.13 shows twin ZnO nanorods grown via a hydrothermal route from a central seed. Careful inspection of the first-order reflections show double spots, indicating that, although the rods grow in the same direction, they have a slight twist from each other (Palumbo *et al* 2008).

As the electrons traverse the sample, they can undergo inelastic scattering and lose amounts of energy specific to the energy level separation of the specific atoms

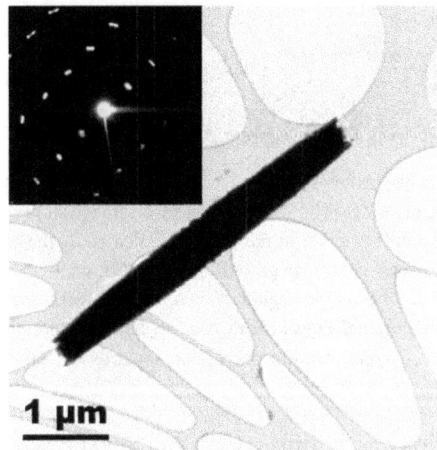

Figure 2.13. A bright-field image of a twin ZnO nanorod, with the rods growing out from a central seed, with the associated diffraction pattern as an inset. The growth direction can be determined from the diffraction pattern as [0001].

within the sample. These electrons can be filtered with an energy-loss imaging filter/ spectrometer and used to form images, called energy-filtered TEM, which show the distribution of elements. This mode of imaging, together with EDX mapping, will be elaborated on in the next chapter.

2.4.4 Precession diffraction patterns—kinematical diffraction

One of the major drawbacks of diffraction patterns in a TEM, in comparison with x-ray diffraction techniques, is that they are dynamical diffraction patterns; this means that the intensities of the peaks change due to interference and cannot be used to relate to the structure factor. However, the TEM can be used for obtaining structural information from small, sub-micron particles, which x-ray diffraction techniques can only access if they are in relatively large quantities, such as powders. In a relatively similar approach to the way Z-contrast imaging averages over several diffracted beams to cancel out the dynamical scattering, Vincent and Midgley (1994) have shown that precessing the beam in a conical pattern (and then descanning to obtain a stationary pattern) results in kinematic diffraction patterns which can be used to resolve the structures of nanoparticles. This technique is commercially available under the NanoMegas trade name and can resolve structures as small as 2 nm in aberration-corrected instruments (Viladot *et al* 2013), including creating orientation maps (similar to the EBSD available in scanning electron microscopes) and strain maps in crystals (Rauch and Veron 2014).

2.4.5 Lorentz lenses and off-axis holography

We have seen in section 2.1 the importance of phase contrast in high-resolution images. At the opposite end of the scale, permanent magnetic and electrostatic fields can also change the phase of the beam; after all, we are using electric and magnetic fields to alter the path of the beam. For magnetic fields, this manifests itself as Lorentz forces, hence the name given to a variant of the objective lens, the Lorentz lens. This is a mini-lens that sits inside the objective lens of the microscope and has very long focal lengths, such that the contrast in the images recorded reveals magnetic domains and their orientations (Lloyd *et al* 2002).

In off-axis holography, a thin conducting wire is negatively charged and placed in the path of the beam, effectively splitting it into two. One beam passes through the sample and has a phase change following interaction, whilst the original beam serves as the reference beam. The two beams are recombined to form an interference pattern, superimposed on the image of the sample, from which we can recover both the amplitude and the phase, as originally described by Gabor (1948). This technique is also useful in measuring magnetic domains, magnetic and electrostatic fields and surface potentials, in conjunction with the Lorentz lens. To study the magnetic behaviour of materials, accurate knowledge of the magnetic fields of the objective lens, as a function of its excitation is necessary (e.g. by calibration with a Hall probe), so that magnetization curves can be followed. The Lorentz lens is a small lens that allows low-resolution imaging with the objective lens switched off; this is because at normal imaging conditions, the magnetic field of the objective lens

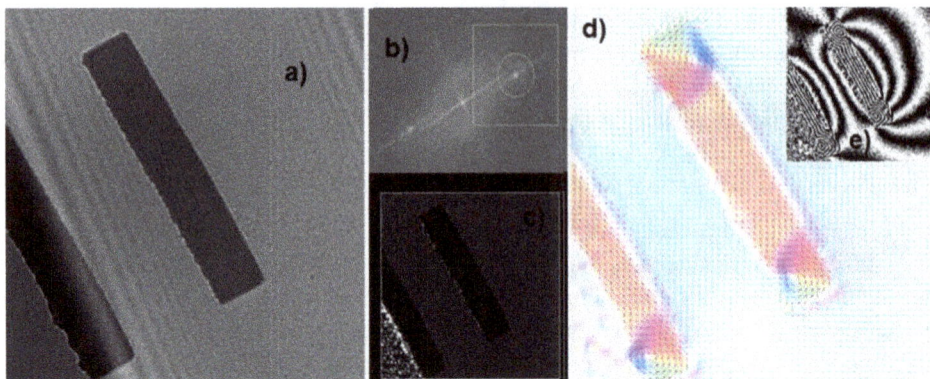

Figure 2.14. (*a*) An interferogram of a magnetic track permalloy element, where small shifts in the interference pattern are phase changes caused by the magnetic field of the permalloy. (*b*) The Fourier transform of (*a*) shows the phase information from the interference pattern as a pass band and we can reconstruct the amplitude in (*c*) and the phase (represented in (e) as $\cos(2\varphi)$), from which we can derive the distribution and orientation of the magnetic field in (*d*). For this aspect ratio, the central portion is the typical single domain magnet, whilst the ends are circular magnet domains.

saturates most materials. Figure 2.14 shows a permalloy element, whose aspect ratio decides the distribution of magnetic domains. The permalloy element is seen with the interference pattern recorded at the same time and is processed in reciprocal space, where the amplitude and the information carried by the interference pattern are easily separable (remember the Lichte defocus discussed in (2.7) and figure 2.3) into amplitude and phase. The phase can be then traced back to the direction of the magnetic field (Dunin-Borkowski *et al* 2000) and can also be represented as $\cos(n\varphi)$, where n is an integer.

2.4.6 3D reconstruction

TEM images are 2D projections of 3D objects and it is possible to reconstruct the 3D sample from a series of images taken at different sample orientations with respect to the beam—a tilt series. This is how medical tomography works as well, with 64 or 128 detectors arranged around the patient, each recording the image at a slightly different tilt. Of note here is that the reconstruction happens in reciprocal space because it reduces the reconstruction problem from 3D to solving in 2D (e.g. a series of 2D planes is represented by a 1D series of spots—a line—in reciprocal space). The draw-back is that the Fourier space is not sampled equally radially and some form of interpolation is required between the experimental data points, which is trickier at higher spatial frequencies, where the high resolution information is. Furthermore, even with a dedicated tomography holder allowing for tilt angles from $-70°$ to $70°$, there is a missing wedge of data which should also be interpolated. Lastly, the thickness of the specimen increases greatly when tilted to such high angles, affecting the contrast and making interpretation difficult. Nevertheless, a considerable body of work now exists, including automated collection and iterative volume reconstructive procedures that work well, provided the sample can withstand the long exposures during a tilt series.

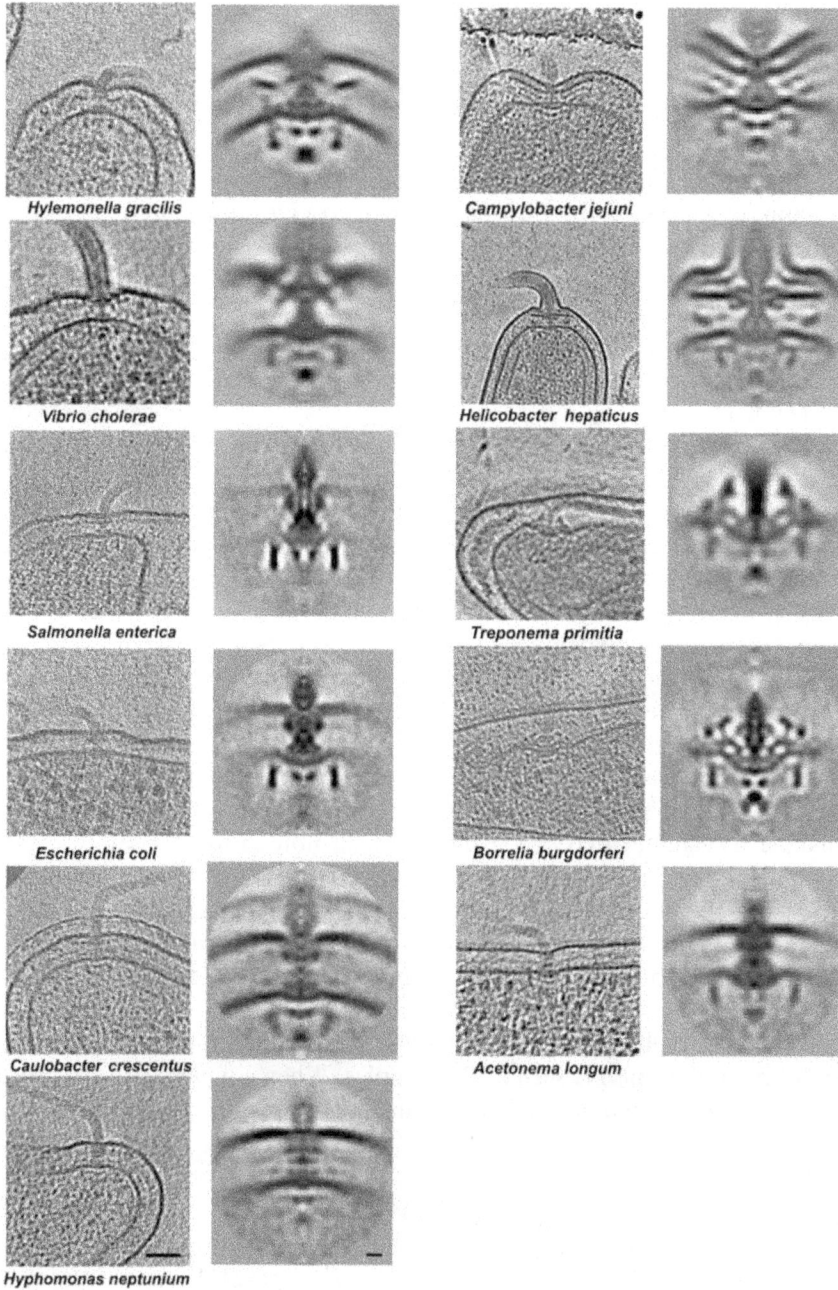

Figure 2.15. Flagellar motor structures obtained by electron cryo-tomography and subtomogram averaging. Left column: 20 nm thick central slices through tomograms of individual cells exhibiting flagellar motors, arranged in the same order as they appear on the phylogenetic tree. Scale bar, 50 nm. Right column: Axial slices through average reconstructions of each motor. Scale bar, 10 nm. The core of the motor remains the same, but the peripheral components vary in complexity, which may be related to the amount of torque these nanomachines can exert. Image reproduced with kind permission from (Chen *et al* 2011). Copyright 2011 John Wiley and Sons, Inc.

One of the applications of this technique was to reconstruct the iron oxide nanoparticles that form the spine of magnetotactic bacteria and show that they were different to the particles that were attributed to organic life forms found in Mars meteorites (Dunin-Borkowski *et al* 1998).

However, it is in life sciences where tomographic reconstruction has made a significant impact, shedding light on the mechanism of protein assembly in the human immunodeficiency virus with sub-nanometre resolution, for example (Schur *et al* 2015). What is striking is that cryo-electron tomography was carried out not just on one virus within a sample, but on hundreds of viruses, which were then averaged (using sub-tomogram averaging (Schur *et al* 2013)) to reveal the structure with sub-nanometre resolution. Automated batch acquisition of tomographic series and reliable reconstruction software make this possible, whilst maintaining the dose low, to below 60 e/Å^2. The same technique was also used by (Chen *et al* 2011) to show the evolution of bacterial motors which help propel them through a medium and relate them to specific genes, as shown in figure 2.15.

It should be only a matter of time before the other imaging techniques commonly used in material sciences (HAADF, holography) also begin to make an impact in the life sciences applications of electron microscopy. Nevertheless, the current applications demonstrated in life sciences show the significant potential impact of electron microscopy as serious competitors to the very expensive synchrotrons.

With the advent of aberration correctors, the probe in the STEM can now go to much higher convergence angles, increasing from a ~15 mrad semi-angle to a ~60 mrad semi-angle, resulting in a significant decrease in the depth-of-field (the thickness of sample that is in acceptable focus). This means that confocal techniques become available and 'slices' as thin as 2–3 nm can now be imaged in Z-contrast mode, such that the 3D volume can be reconstructed much more easily (Borisevich *et al* 2006).

In summary, we have discussed some of the significant concepts around the resolution of the microscope, differentiating between the information limit and the point resolution. We have provided an insight into image calibration, but also the importance of alignment of the sample with respect to the beam direction. We have then briefly reviewed imaging techniques, including diffraction, and have expanded on the usefulness of several signals available at the same time in the interpretation of the changes seen in the samples. When coupled with spectroscopy, TEM and STEM become unrivalled instruments for analysing nanomaterials.

References

Adams A R, Marko I P, Mukherjee J, Stolojan V, Sweeney S J, Gocalinska A, Pelucchi E, Thomas K and Corbett B 2015 Semiconductor quantum well lasers with a temperature-insensitive threshold current *IEEE J. Sel. Top. Quantum Electron.* **21** 6

Allen L J, Findlay S D, Oxley M P and Rossouw C J 2003 Lattice-resolution contrast from a focused coherent electron probe. Part I *Ultramicroscopy* **96** 47–63

Borisevich A Y, Lupini A R and Pennycook S J 2006 Depth sectioning with the aberration-corrected scanning transmission electron microscope *Proc. Natl Acad. Sci. USA* **103** 19212

Chen S Y *et al* 2011 Structural diversity of bacterial flagellar motors *EMBO J.* **30** 2972–81

Crewe A V 1966 Scanning electron microscopes – is high resolution possible? *Science* **154** 729

Crewe A V 1971 High resolution scanning microscopy of biological specimens *Phil. Trans. R. Soc.* B **261** 61–70

Crewe A V, Wall J and Welter L M 1968 A high-resolution scanning transmission electron microscope *J. Appl. Phys.* **39** 5861

Dunin-Borkowski R E, McCartney M R, Frankel R B, Bazylinski D A, Posfai M and Buseck P R 1998 Magnetic microstructure of magnetotactic bacteria by electron holography *Science* **282** 1868–70

Dunin-Borkowski R E, McCartney M R, Kardynal B, Parkin S S P, Scheinfein M R and Smith D J 2000 Off-axis electron holography of patterned magnetic nanostructures *J. Microsc.* **200** 187–205

Gabor D 1948 A new microscopic principle *Nature* **161** 777–8

Giusca C E, Stolojan V, Sloan J, Boerrnert F, Shiozawa H, Sader K, Ruemmeli M H, Buechner B and Silva S R P 2013 Confined crystals of the smallest phase-change material *Nano Lett.* **13** 4020–7

Hashimoto H, Endoh H, Tanji T, Ono A and Watanabe E 1977 Direct observation of fine-structure within images of atoms in crystals by transmission electron-microscopy *J. Phys. Soc. Japan* **42** 1073–4

Lehmann M and Lichte H 2002 Tutorial on off-axis electron holography *Microsc. Microanal.* **8** 447–66

Lloyd S J, Loudon J C and Midgley P A 2002 Measurement of magnetic domain wall width using energy-filtered Fresnel images *J. Microsc.* **207** 118–28

Palumbo M, Henley S J, Lutz T, Stolojan V and Silva S R P 2008 A fast sonochemical approach for the synthesis of solution processable ZnO rods *J. Appl. Phys.* **104** 074906

Rauch E F and Veron M 2014 Automated crystal orientation and phase mapping in TEM *Mater. Charact.* **98** 1–9

Scherzer O 1949 The theoretical resolution limit of the electron microscope *J. Appl. Phys.* **20** 20–9

Schur F K M, Hagen W J H, De Marco A and Briggs J A G 2013 Determination of protein structure at 85 angstrom resolution using cryo-electron tomography and sub-tomogram averaging *J. Struct. Bio.* **184** 394–400

Schur F K M, Hagen W J H, Rumlova M, Ruml T, Muller B, Krausslich H G and Briggs J A G 2015 Structure of the immature HIV-1 capsid in intact virus particles at 88 angstrom resolution *Nature* **517** 505

Shibata N, Pennycook S J, Gosnell T R, Painter G S, Shelton W A and Becher P F 2004 Observation of rare-earth segregation in silicon nitride ceramics at subnanometre dimensions *Nature* **428** 730–3

Sidorov M 2002 ctfExplorer http://www.maxsidorov.com/ctfexplorer/

Viladot D, Veron M, Gemmi M, Peiro F, Portillo J, Estrade S, Mendoza J, Llorca-Isern N and Nicolopoulos S 2013 Orientation and phase mapping in the transmission electron microscope using precession-assisted diffraction spot recognition: state-of-the-art results *J. Microsc.* **252** 23–34

Vincent R and Midgley P A 1994 Double conical beam-rocking system for measurement of integrated electron-diffraction intensities *Ultramicroscopy* **53** 271–82

Williams D B and Carter C B 2009 *Transmission Electron Microscopy: A Textbook for Materials Science* (New York: Springer)

Chapter 3

Spectroscopy

The 'colour' of the electrons that have scattered inelastically forward from the sample can be determined by measuring the electrons' energies, using an EELS. This offers sub-nanometre spatially resolved information about the chemical nature of the sample, the unoccupied electronic density of states and the density of electrons. Together with the imaging techniques described in the previous chapter, it is a significant tool in understanding the processes and phenomena on the nanoscale. At the same time, the energy transferred to the sample can be re-emitted as characteristic x-rays, which can also be collected with an EDX spectrometer, offering compositional information about the sample with high spatial resolution. Other spectroscopic signals are available, such as cathodoluminescence and Auger-electron spectroscopy, but here we will only discuss EDX and EELS.

As each transition occurs between energy levels which are specific to each atomic element, the composition of a sample can be determined. In general, EELS covers energy-loss ranges from 0–2000 eV, beyond which the electron energy-loss (EEL) signal becomes very low. On the other hand, EDX can cover a wide range, being able to detect x-rays with energies up to 40 keV (depending on the spectrometer). In general, one uses EELS for mapping and quantifying light elements (particularly C, N and O), including 3d transition metals and EDX is generally used for heavier elements (e.g. $Z > 10$). Whilst the spectral resolution of EELS (in energy) is of the order of meV and is limited by the energy spread of the electron source, EDX spectrometers generally have ~120 eV energy resolution, which makes separating signals of various elements, particularly oxides of transition metals and carbides and nitrides, more complicated. The acquisition times also vary significantly, with EDX generally one to two orders of magnitude longer than EELS, but recent advances in EDX detector technology have shaved one order of magnitude from the typical exposure times. It is possible to improve the resolution of x-ray spectroscopy by switching to wavelength-dispersive spectroscopy (~15 eV), but this is a more complex and slower technique, where the wavelengths are scanned serially, rather than in parallel, as with the EDX. However, the detectability limit and the accuracy in composition improve by at least an order of magnitude.

doi:10.1088/978-1-6817-4120-8ch3

With both spectroscopies, performing chemical composition analysis relies on good statistics for the signals from all component elements being present. All the elements which are present in a given spectrum are considered, for analysis, to add up to 100% and elements below the detectability limit or outside of the energy range of the spectrum are disregarded. This means that, in some cases, we consider relative ratios of one element to the majority element in the sample, rather than absolute atomic or weight % compositions.

The spatial resolution of the EELS signal is defined by the beam size and the spatial extent of the event excited and, as a rule-of-thumb, spreads by a factor proportional to the thickness, t, of the sample. Similarly, the EDX signal spreads by a factor proportional to $t^{3/2}$. The thicker a sample is, the more likely it is that at least one inelastic scattering event will occur as an electron traverses the sample (higher intensity signal—for EDX, but not for EELS—see section 3.2), but also the worse the spatial resolution becomes.

For both techniques, the propagation of the beam through the sample can have important effects, particularly when the electrons travel along a strongly diffracting direction in crystalline samples. This leads to strong dynamic effects, such as oscillations in the intensity of the electron beam as it travels down the column of atoms. Normally these conditions are avoided by tilting away from strong diffraction conditions, which is trickier when looking at information across interfaces (see figure 2.8). In particular cases, channelling can be used to enhance microanalysis in both EDX (Spence and Tafto 1983) and EELS (Kirkland 2005, Saito *et al* 2009).

3.1 EDX

In x-ray spectroscopy, the high-energy electron can excite a deep core electron to unoccupied states in the outer shells. Upon de-excitation, energy is emitted as a photon in the x-ray range. As there are several routes for an electron to recombine with the deep core hole, we usually observe a series of lines corresponding to several electron de-excitation pathways. The nomenclature of x-rays is relatively straightforward: <u>a capital letter</u>, indicating the shell with the initial vacancy/hole that the electrons de-excites to (K: $l = 0$; L: $l = 1$; M: $l = 2$; l = orbital angular momentum) followed by <u>a Greek letter</u> which indicates the energy level that the electron de-excited from (α: $\Delta l = 1$; β: $\Delta l = 2$, etc). For example, the Cu Kα line represents the x-ray emitted by an atom of Cu when the electron de-excites to the 1s (K) energy level from the 2p ($\Delta l = 1$) energy level; if we want to distinguish between the spin orbit splitting (remembering that the typical energy resolution of an EDX spectrometer is 123 eV), then Kα_1 represents transitions from the $2p_{3/2}$ higher energy level and Kα_2 transitions from the lower $2p_{1/2}$ energy level. A typical EDX spectrum (figure 3.1) shows a series of peaks, displayed as a function of energy. The area under a peak is proportional to the amount of that particular atom in the sample and the cross-section of the respective x-ray emission process for that specific atom.

Modern spectrometers have a high amount of automation integrated in their software platforms, with most spectrum processing and analysis happening in the background, and results usually presented in report format or database tables. Peak processing, fitting, separation and identification is nowadays performed rapidly and

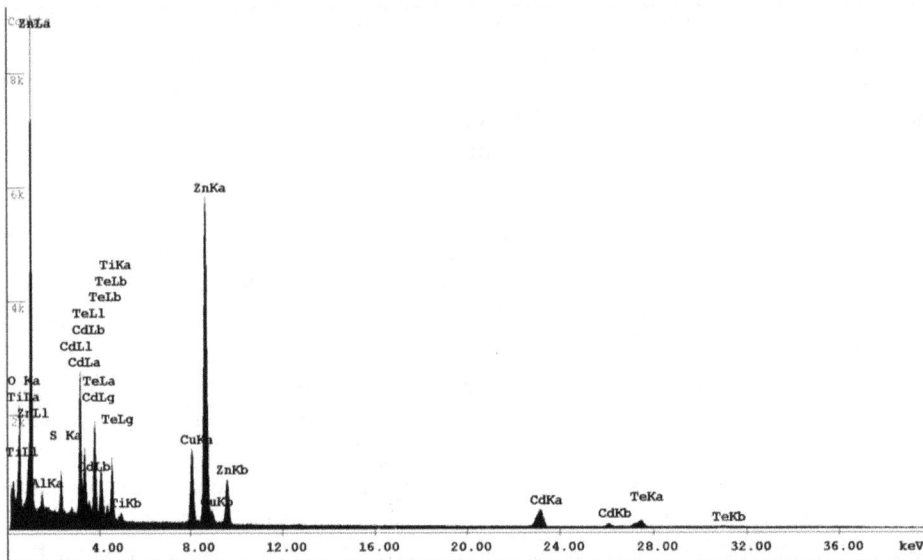

Figure 3.1. A typical EDX spectrum from the solar cell sample in figure 2.11, with the peaks assigned to the respective atomic elements. The Ti and Cu signals in this case come from the sample holder. The identification of the elemental L series, which overlap each other, is helped by the $K\alpha$ lines, which have larger separation in energy and are more easily distinguished.

in real-time, even for relatively low signal-to-noise ratios. Nevertheless, operator input is still necessary to discriminate spurious results, the correction procedures applied and the methods of calculating the elemental ratios.

Whether the sample analysis is qualitative (e.g. elemental distribution maps) or quantitative (e.g. area or point-by-point compositional analysis), the goal is to obtain as many counts as possible. This is a balance between radiation damage, beam intensity and acquisition time. Within the total acquisition time (how long the beam is on the sample for) there are two components: the live acquisition time (how long the spectrometer is acquiring for) and the processing 'dead' time (when the spectrometer is discriminating the x-rays recorded). The longer the processing time, the better the energy resolution in the spectra and the more accurate the discrimination between close peaks, such as the lower Z elements, but the longer the total acquisition time needed. No EDX, and generally no spectroscopy, should be performed without first testing the resilience of the sample and the contamination during long exposures, which can range up to the order of tens of minutes for EDX.

EDX spectrometers rely on calibration standards, samples of known composition which can be used to calibrate the energy scale of the spectrometers, by measuring the energy separation of two known lines in the same spectrum, e.g. Cu and Al. This calibration should be performed regularly and cover all energy dispersions and all acquisition time constants available for a specific spectrometer. This ensures that the peak energies and their subsequent assignment to specific elements is correct. For quantitative compositional analysis, the use of standard samples is necessary to relate the known compositions to the relative peak intensities, essential when trying to analyse compounds with unknown stoichiometry.

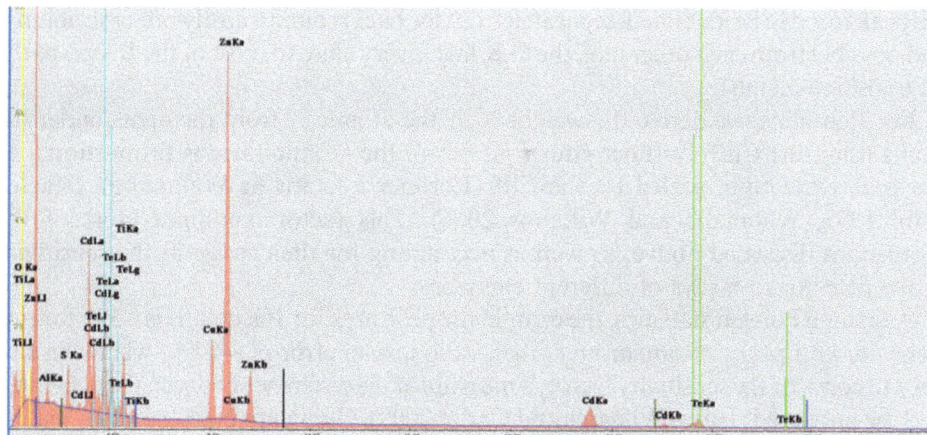

Figure 3.2. A typical screenshot of an EDX spectrum, with a (linear) background shown together with library data for tellurium, indicated where peaks should be expected, including the absorption, sum and escape peaks.

One of the challenges related to EDX is understanding peak artefacts in the spectrum. The x-rays, upon being emitted, can undergo further interactions such as scattering from the sample holder and the microscope components, including exciting x-rays in the Si of the detector. All these effects range from adding to the background (bremsstrahlung, when a charged particle—the incident electron—slows down without exchange of energy to an atomic transition it emits x-rays in a continuous spectrum) to resulting in artefact peaks, such as the Cu and Ti peaks seen in figure 3.1 which originate from the components of the sample holder. Other artefact peaks are sum peaks (emitted x-rays are absorbed and re-emitted by another atom species), escape peaks (x-rays which are not absorbed in the detector, but escape as Si x-rays so the total energy is shifted down—1.74 keV for Si detectors) or coherent bremsstrahlung peaks (due to the very thin samples, the bremsstrahlung peaks are Gaussian-shaped). In general, most analytical EDX software will have an option to display these peaks (figure 3.2), as well as offering a range of possible elements that can give rise to that peak, within the spectral resolution. In most cases though, it is the experimentalist that needs to make a decision on which peak is which, either by matching several peaks to one element or, if the peaks are too close to other transition energies, to confirm the presence of that particular atomic species by other techniques, such as EELS. In the case of trace elements, good statistics become the most important aspect of quantification and confidence levels are directly related not just to the number of counts, but also to the signal-to-noise ratios and the signal-to-background ratios. With EDX, the unexpected presence of an element in the spectrum should be generally verified with other techniques, e.g. XPS, particularly if it appears at the trace level.

Once the peaks are identified, a number of processing steps need to be performed before quantification, most of which are automated in all manufacturer proprietary software. Still, the user must be aware of the corrections that need to be applied to the spectra for the case of thin samples irradiated with high energy electrons, such as absorption (preferential x-ray absorption by specific elements) and fluorescence (emission of secondary x-rays, following primary x-ray absorption), particularly in comparison to EDX in scanning electron microscopes. Before measuring peak areas, the peaks need to be identified, separated from the background (mostly bremsstrahlung) and possibly from each other (e.g. the O K line is very close to most of the L lines of the 3d transition metals).

For thin films, we derive the weight % or the atomic % from the areas under the peaks using the Cliff–Lorimer equation, where the atomic ratio is proportional to the peak area ratio, scaled by the Cliff–Lorimer k-factor or Watanabe's ζ-factor (Eibl 1993, Watanabe and Williams 2006). This factor combines most of the corrections discussed above, as well as accounting for the change in the sensitivity of the detector to x-rays of different energies.

Assuming Poisson statistics, the error in the peak area for 10k counts is ~3%; for two elements with peaks of similar area, their ratio has an error of ~4.5%, where we also need to consider the sensitivity factor, a measure of the difference between the generated and the measured x-rays (Williams and Carter 2009). Going up to 100k counts, readily accessible with the new generation of solid-state detectors, the error of a ratio of two peaks drops to ~2%. However, remember that the thinner the sample, the fewer the counts, so attempting to extract EDX counts from a 2 nm sized catalyst in a nanowire/nanotube may require tens of minutes of acquisition time and a sample that withstands that type of irradiation damage, not to mention a good drift correction routine!

The accuracy in calculating the k-factor really comes down to using a suitable standard, one of a known composition and that is stable under the beam, homogenous and without contaminants. Nevertheless, for a reliable, experimental estimate of errors in quantification, one should always repeat the experiment several times and, perhaps, over several areas.

The detectability limit, the smallest amount of a material that can be detected, is related to the spatial resolution and the volume of the sample that generates x-rays (sample thickness, probe size, probe current). In other words, 0.1 wt% may be possible for a thicker sample and lower resolution (Williams and Carter 2009), whilst 0.5–1 wt% is achievable with sub-nanometre resolution, aberration-correction and the new generation of solid-state x-ray detectors (Lewis *et al* 2014).

As with any scanning technique, with the EDX in acquisition mode, the electron probe can be rastered across an area, in a series of spots across a line (e.g. across an interface) or held in a spot for the duration of the acquisition, leading to 2D maps, line profiles, or single-spot spectra respectively. One of the issues with area and line scanning is that the user must consider the dimensions of the probe and the interaction volume versus the area of the 'pixel' in an area scan or a line scan. Usually, the size of the probe is smaller than the size of the pixel, which means there is uncertainty where the spot dwells in relation to the pixel itself, which is significant when crossing an atomically sharp interface for a very thin sample—one can effectively miss the target. This is usually solved by using sub-pixel scanning, a feature currently available only in Digital Micrograph™ and derived from EELS acquisition strategies. Perhaps what is missing at this moment for specific cases with planar interfaces, such as between functional layers of semiconductor devices, is the ability to step a line-shaped probe across the interface; the line shape could be formed by scanning the circular probe backwards and forwards parallel to the interface during the acquisition of one spectrum, reducing damage and contamination. Interfaces are generally less stable regions, with defects and higher concentration of impurities and would be more affected by beam damage, such as the introduction or annihilation of defects, impurity diffusion and sputtering hence, reducing the dose is beneficial.

Qualitative EDX is usually used to acquire 2D elemental maps in scanning mode, where a number of integrating windows are set up and the values of those integrals are displayed at each of the scanned positions, building maps of the distribution of specific elements that also show relative changes in concentration for a single element. The mapping is often referred to as semi-quantitative; as the area maps also contain the background—full quantisation requires this to be removed.

Figure 3.3 shows a series of maps, together with the simultaneously acquired Z-contrast image, of a FIB section of a solid-oxide fuel cell, showing the area of the

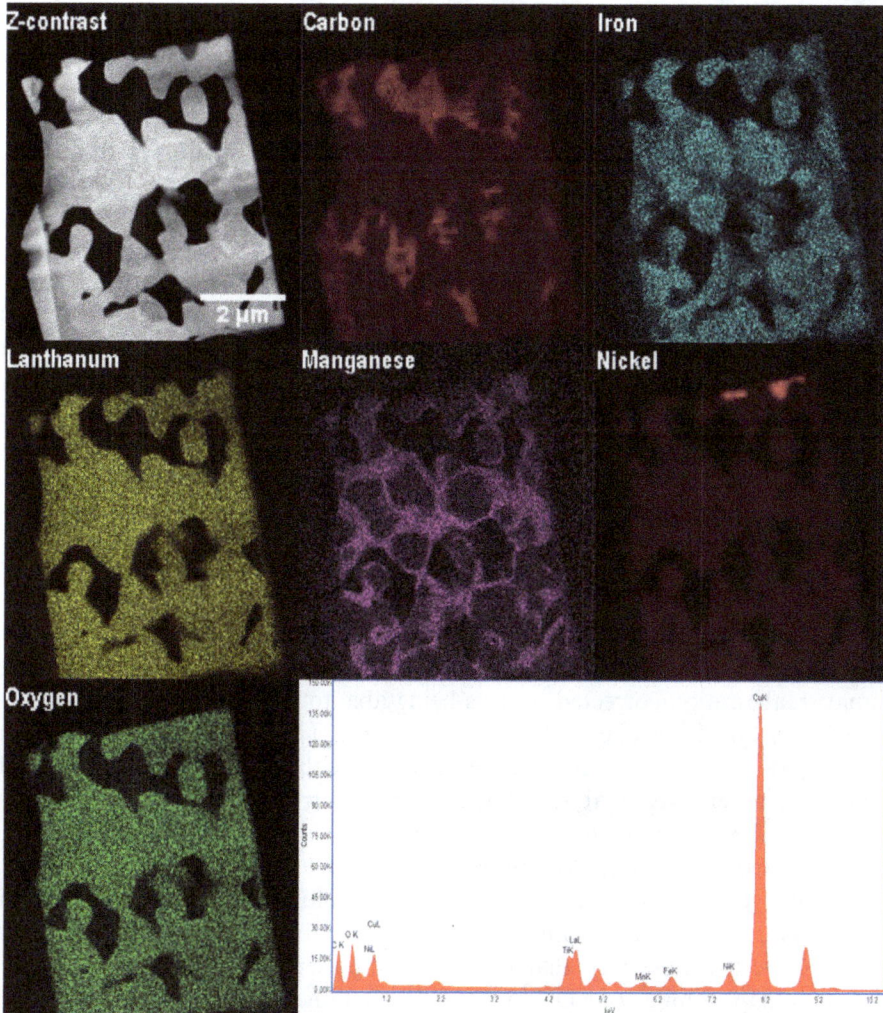

Figure 3.3. A set of EDX elemental maps for a FIB-cut foil, supported on a holey-carbon film, of a LaNi$_{0.6}$Fe$_{0.4}$O$_3$ current-collector layer in a solid-oxide fuel cell, showing that there is no zirconium lanthanate product, but that the Ni, Fe and Mn diffuse, with Fe-rich grains and Mn segregated to the grain boundaries. As Ni diffuses out, grain boundaries are blocked by Mn, ultimately leading to the failure of the fuel cell.

lanthanum oxide current-collector layer, with the cathode (not shown) at the top of the picture. The FIB section is supported on a holey-carbon grid, which is visible through the holes in the perovskite current collector (see the carbon map). In the $LaNi_{0.6}Fe_{0.4}O_3$ (the current collector) we note that the iron is in the central part of the grains, whilst the manganese has segregated to the grain boundaries. From these maps, we conclude that the diffusion of the Mn and Ni from the cathode current-collector layer to the yttria-stabilised zirconia cathode is responsible for the decrease in solid-oxide fuel cell performance (Millar *et al* 2008).

Alternatively, qualitative information can be extracted by performing line profiles across features of interest, such as across interfaces in cross-sectional layers, as seen in figure 2.11, and techniques such as concentration histogram analysis can be used to extract the phase composition of a sample and the spatial distribution of the phases (Bright and Newbury 1991).

Quantitative mapping requires the collection of the full spectrum, as well as fitting and separation of the peaks from the background. Acquiring this information over an area can become a very lengthy process, since each acquisition point requires exposures of several seconds (or longer, for good statistics or trace elements). If the sample can withstand this amount of irradiation, then the remaining requirement is that sample drift is accounted for regularly, either automatically or manually, using a fiducial marker.

EDX is relatively simple to use, is efficient for a wide range of elements and is comparatively less expensive than EELS. However, it has a worse spatial and energy resolution and is not really suited to light elements, particularly when several are present (e.g. B, C, N and O), whose K-edges are quite close together (188 eV, 285 eV, 401 eV and 532 eV, respectively), very close to the spectrometer's energy resolution. The two techniques combined, however, transform the electron microscope into an incredibly powerful analytical technique, able to provide insight into nanoscale processes and phenomena in materials and devices.

3.2 EELS

EELS can resolve energy transitions down to vibrational spectroscopies, with mono-chromated, aberration-corrected sources being able to achieve energy resolutions of ~10 meV (~1 nm size probe) (Krivanek *et al* 2014). This brings together the spatial resolution[1] of the electron microscope together with the energy resolution of a synchrotron, as envisaged at the dawn of the aberration correction era (Brown 1997). The EELS can serve not only to measure the electronic structure of the sample, but can also filter inelastically scattered electrons out, leading to improvements in high-resolution images, the sharpness of spots and high-resolution information in diffraction patterns and in the contrast of images from thicker samples. This filtering can occur also at specific energies, such that images can be then reformed to give elemental distribution maps, similar to EDX in figure 3.1, but one element at a time (serially).

---

[1] The reader is forewarned at this stage that spatial resolution for EELS combines both the probe size (which would limit the resolution in Z-contrast imaging) and the spatial (de)localisation of the inelastic event, which will be discussed further on.

In EELS, we can measure the energy lost by an electron through inelastic interaction with the sample either by using one or more magnetic prisms (a magnetic sector), or by using a combination of magnetic and electrostatic fields (a Wien filter). In both cases, we use the Lorentz force (3.1) to deflect the electrons more or less (in a spectrometer, B is normal to the velocity, v), depending on their velocity. For the Wien filter, we also use an electrostatic field (operating perpendicular to the magnetic field, both normal to the electron velocity), which selects only electrons of a particular velocity to continue on the same path as the incoming electrons. This force acts as a centripetal force, leading to a curved trajectory of the scattered electron, the radius of which (and therefore the exit direction, i.e. energy dispersion) depends on its velocity.

$$f_{\text{Lorentz}} = -e \, v \times B. \tag{3.1}$$

The Wien filter has found use in monochromating the electron beam in an aberration corrected microscope, by placing a small slit at the exit to select a small energy range of electrons. It is usually located before accelerating the electrons to the high voltage of the microscope and can be combined with corrective and focusing elements to lead to very narrow energy spreads (Tiemeijer 1999, Krivanek *et al* 2013).

The market for magnetic sector spectrometers is dominated by Gatan Inc., and is associated with specialised TEM image acquisition and processing software (Digital Micrograph™), as well as the integrated EELS acquisition and processing routines. After dispersion by the spectrometer (a typical imaging and spectroscopy instrument found on TEMs), we can describe the information as a 3D data set (figure 3.4), where two axes are the spatial co-ordinates of the projected image of the sample and the third axis is the energy loss. We can then describe the EELS as manipulating this data set

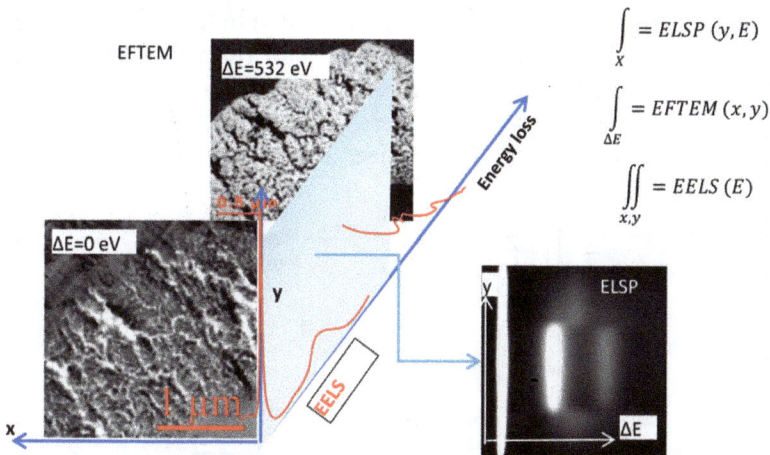

Figure 3.4. Illustration of the 3D data set available with a typical Gatan Imaging Filter EELS, suitable for TEM. We can use an energy selecting slit to form images with electrons that have lost a specific range of energies by integrating the 3D data set over an energy interval, ΔE. To obtain a spectrum, the spectrometer projects the 3D data set along one of the spatial co-ordinates, in this case x, onto a 2D detector (i.e. integrate over x), which is then binned (i.e. integrate over y) to obtain the EEL spectrum. The intermediate case (integrate over x but not y) is called ELSP. The two energy-filtered images shown at 0 eV and 532 eV (O K-edge) energy loss are the Al–Al$_2$O$_3$–epoxy sample discussed later on in figure 3.9, whilst the ELSP image is from a layer of Al (middle) on Si.

(rotation, sectioning in 2D) so that it can be projected onto the detector, a yttrium–aluminium–garnet scintillator bonded to a Peltier-cooled charge-coupled device, or the more novel direct electron detector, based on complementary metal-oxide semiconductor technology. For example, rotating the 3D data set so that one of the position axes is normal to the detector results in an image which retains one spatial co-ordinate and the energy co-ordinate. This technique is called energy-loss spectroscopic profiling (ELSP) (Walther 2003). For samples with linear features, such as interfaces, they can be aligned normal to the y-axis (e.g. with a tilt-rotate holder, or by using one of the magnetic lenses to rotate the image) resulting in a parallel acquisition of spectral information across the interface (figure 3.4). This is the closest to having a line-shaped probe that we can scan across the interface, reducing the dose and damage. Note that the length of the line probe is not constant, but defined by the collector aperture width along the vertical axis. However, all positions across the interface are recorded simultaneously, which means that they are recorded on the same energy scale (see core-level shifts in section 3.2.5). Binning this image along the y-axis results in a 1D typical EEL spectrum, as can be seen in figure 3.5.

STEM-specialised spectrometers do not usually have the imaging hardware and are optimised for fast acquisition, binning and data recording, as well as being able to cope better with the high dynamical range present in an EEL spectrum.

3.2.1 The EEL spectrum

A typical EEL spectrum is displayed in figure 3.5 and is described by three regions: the zero-loss peak (ZLP), the low-energy-loss region (0–100 eV) and the high-loss region (>100 eV). A thin sample in terms of EELS is generally defined as one where the ZLP is at least 10 times higher than the height of the next highest peak, which is the bulk plasmon resonance peak. This ratio is related to the thickness of the sample in terms of being smaller than the mean free path (mfp) for inelastic scattering. For the experiment, the user controls the exposure time and the collector aperture, both of which can be used to reduce or increase the recorded intensity, to cope with the huge typical dynamic range. In practice, an electrostatic beam blanker is also

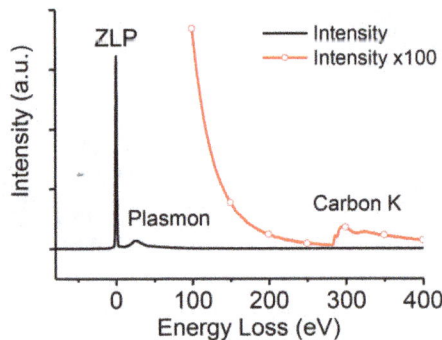

Figure 3.5. A typical EEL spectrum from an amorphous carbon film, showing the three regions of interest: the ZLP, the plasmon peak and the carbon K-edge. To reveal the carbon K-edge, the intensity has been increased by a factor of ~100 by increasing the exposure time from tens of milliseconds to seconds and increasing the acceptance aperture.

necessary to achieve very short exposure times, as changing the collector aperture to reduce the exposure for the ZLP (the spectrometer entrance aperture) will, as we will see further, have significant effects on the data collection and processing.

The interaction of the fast electron with the sample can either be described via electrodynamics, where the electron creates a time-varying field, polarising the material (described by a complex dielectric function) and slowing the electron down, or by using the quantum mechanical description of the transition rate theory, as the probability of scattering from an initial state i to a final state f. It is common to use the dielectric description for the low-energy-loss part of the spectrum, particularly in relation to collective modes of oscillation (plasmons) and the measurement of the imaginary part of the dielectric function; the real part is then recovered using the Kramers–Kronig relation. The transition rate theory is more commonly used for the description of single-electron events (high-loss).

In the dielectric description, the double-differential scattering cross-section, per atom, describes the probability of the electron losing energy E and being scattered into the solid angle $\Omega = 2\pi\theta\,\mathrm{d}\theta$ and is given by (Ritchie 1957):

$$\frac{\mathrm{d}^2\sigma}{\mathrm{d}\Omega\mathrm{d}E} \approx \frac{1}{\pi^2 a_0 m_0 \nu\, n_a} \left(\frac{1}{\theta^2 + \theta_E^2}\right) \mathrm{Im}\left[-\frac{1}{\varepsilon(\mathbf{q}, E)}\right] \quad (3.2)$$

where a_0 and n_a are the Bohr radius and the number of atoms per unit volume and m_0 and ν are the rest mass and the velocity of the electron. $\mathbf{q}$ is the scattering wave-vector, the difference between the incident and the scattered momenta of the electron.

For forward scattering (low scattering angles[2]), the angular dependence (the Lorentzian second term in (3.2) of the scattering cross section has a half-width of θ_E, which is called the characteristic scattering angle. A rough approximation, for incident energies below 150 keV, for θ_E is $E/(2E_0)$, where E_0 is the incident electron energy. The meaning of this value is that the scattering cross-section is relatively constant (independent of $\mathbf{q}$ and E) for collection angles below θ_E. For the carbon K-edge at 200 keV, this value is 1.43 mrad, whilst typical collection angles used in EELS are an order of magnitude higher. As the collection angle increases from θ_E, the scattering cross-section reduces at first linearly, then more rapidly up to a cut-off angle θ_c, which is known as the Bethe ridge (Egerton 2009), beyond which the single electron peak is reduced in intensity. The implication of this is that using collection angles wider than this cut-off will add very little to the signal, but increase the noise. In practice, the exposure conditions are found by monitoring the so-called *jump ratio*, which is the ratio of the feature signal to the pre-feature background. The dipole selection rule applies for EELS when the collection angle is small and allowed transitions are those where there is a unit change in the angular momentum quantum number ($\Delta l = \pm 1$). As the collection aperture is increased, non-dipole transitions can sometimes be observed (Gloter *et al* 2009).

The third term in (3.2) is called the response function and we specified its dependence on the scattering vector $\mathbf{q}$ to also include the response of anisotropic materials, such as graphite. The carbon K-edge in figure 3.6 shows two distinct

---

[2] Whilst we use 'angle' here for convenience, the correct term is 'semi-angle'.

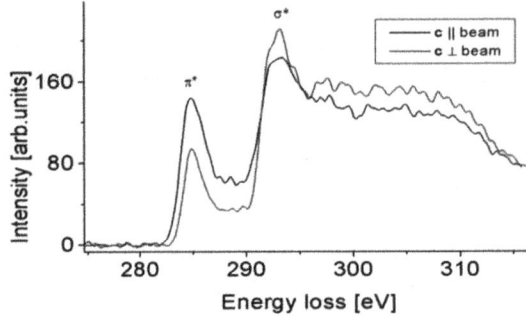

Figure 3.6. The π^* and σ^* carbon K-edge peaks in graphite change in relative heights depending on the orientation of the graphite c-axis in relation to the electron beam. The collection angle at which the parallel and perpendicular components of the dielectric function are weighted equally is the *magic angle*.

peaks, corresponding to the unoccupied sp^2-hybridised orbitals, π^* and σ^*. For amorphous carbon thin films, such as the ones used to fabricate the superlattice in figure 1.4, the amount of sp^2-hybridised carbon relative to the sp^3-hybridised carbon is related to the band-gap and the optical properties of the film (Ferrari *et al* 2000b). By assuming that only sp^2- and sp^3-hybridised carbon bonds are present, we can determine the sp^2-content from the fitted area of the π^* peak normalised to the total area of the π^* and σ^* peaks (Berger *et al* 1988, Papworth *et al* 2000). However, the parallel and perpendicular-to-the-beam components of the randomly oriented sp^2-bonds are not weighed equally, unless the collection angle β is the *magic angle*, $\beta = 2\theta_E$ (Jouffrey *et al* 2004, Menon and Yuan 1998, Daniels *et al* 2003).

An optical spectrum is, by comparison to (3.2), proportional to $\mathrm{Im}[\varepsilon(0,E)]$. For high energy losses, where the imaginary part of the dielectric function ε_2 is small and the real part $\varepsilon_1 \sim 1$ then the EEL spectrum $\mathrm{Im}[-1/\varepsilon(\mathbf{q},E)] \approx \varepsilon_2$, as for the optical response.

If we now look at the transition-rate theory, first developed by Bethe in the 1930s (Bethe 1929), we can write the equivalent of equation (3.2) in the dipole approximation, as:

$$\frac{d^2\sigma}{d\Omega\, dE} \approx \frac{m_0 e^4}{2\varepsilon_0^2 h^2}\left(\frac{1}{E_0 \mathbf{q}^4}\right)|\langle f|r|i\rangle|^2. \qquad (3.3)$$

The transition rate term represents a plane wave transition from the initial state i to the final state f. This has again implied the dipole approximation, where the exponential e^{iqr} term can be expanded into Taylor series and we retain the expansion up to the dipole term.

If we now use the description of the atomic wave function as a linear combination of atomic orbitals, then we can further project[3] the final state onto an atomic orbital, which is the local density of states. In other words, the C K-edge in figure 3.6 is the convolution between the 1s (nearly a delta function, by comparison) and the 2p orbitals and is the unoccupied density of states, symmetry-selected ($\Delta l = \pm 1$) and

---

[3] Mathematically this means a convolution between the two orbitals, initial and final.

site-projected (defined by the size of the probe and the extent/localisation of the final state—e.g. the size of the 2p orbital for the carbon K-edge in figure 3.6). This is why EELS can claim atomic lateral[4] resolution, able to map individual dopants in atoms of columns (Xin *et al* 2014). The advent of sub-atomic-sized probes means that we can now explore the delocalisation effects due to bonding and impurities, allowing us to shed light onto the effects of individual dopants, such as for storing spin states for use in quantum computing (Litvinenko *et al* 2015).

If the initial state is broad/delocalised, such as an interband transition, then the EEL spectrum describes the joint density of states, from which we can recover the optical absorption spectrum.

3.2.2 The low-loss spectrum and plasmon excitations

As stated earlier, the highest intensity peak in the EEL spectrum is the bulk plasmon, which represents energy lost to the collective oscillation of the valence electrons (in metals and some semi-conductors) or the collective response of the induced dipoles in high band-gap semiconductors and insulators. These are delocalised events, extending several nanometers; this can also lead to elastic effects in the spectra, leading to intensity modulation and interference (figure 3.7).

Surface and interface plasmons on the other hand, where the oscillations are constrained in one of the dimensions by the presence of a dielectric interface, can be measured and imaged with nanometer spatial resolution across the interface, since the electric fields of the plasmon waves decay rapidly in the case of the

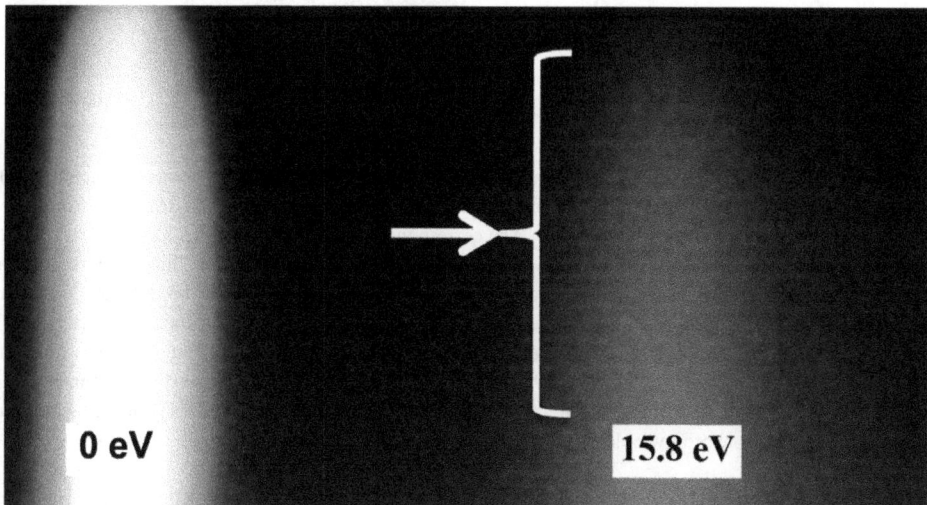

0 eV　　　　　　　　　　　　　　**15.8 eV**

Figure 3.7. Line profile spectra (ELSP) of GaAs showing lattice fringes at the plasmon energy loss (indicated by the arrow-bracket), proving that elastic effects can occur particularly for delocalised interactions, such as plasmons.

---

[4] We use 'lateral' here to remind the reader that the sample, unless graphene or a similar class of material, has a thickness of a few hundred atoms, even though the depth of field of the new aberration corrected microscopes in STEM mode is of the order of nm.

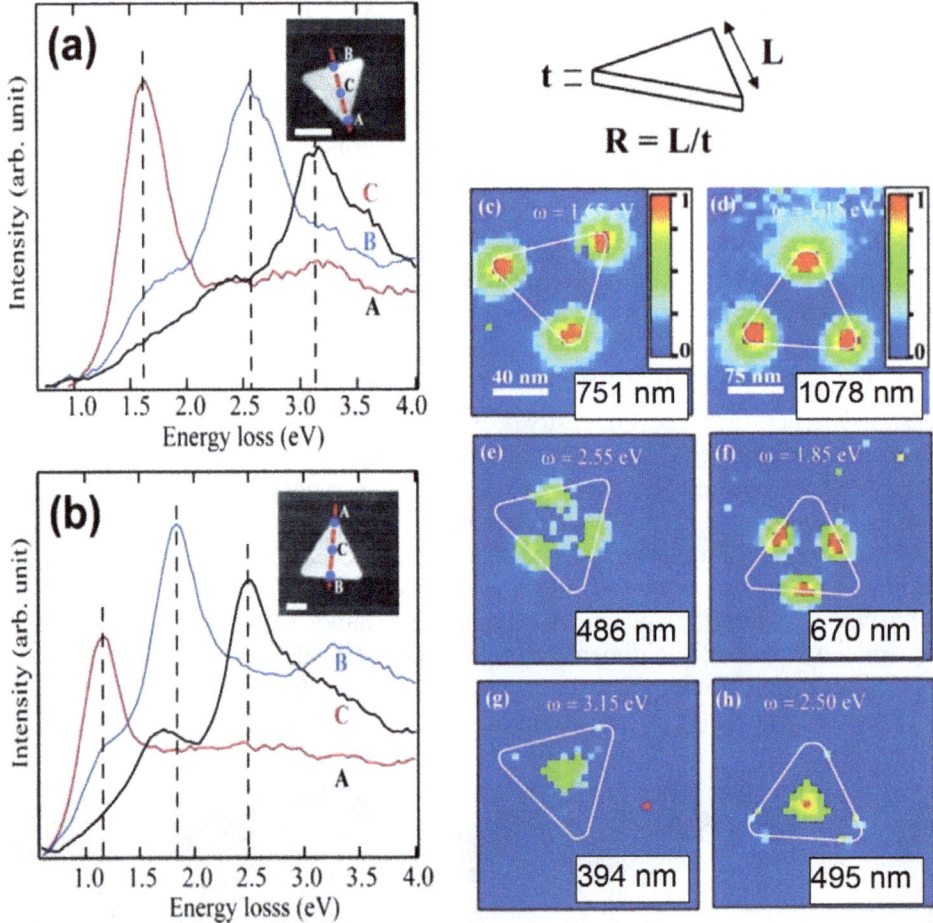

Figure 3.8. Size dependent mapping of plasmons in silver nanoprisms. EEL spectra acquired at a corner (A), the middle point of an edge (B) and the centre (C) of (*a*) 97 nm edge-long (thickness 4 nm) (*b*) and 176 nm edge-long (thickness 6 nm) nanoprisms, respectively. The corresponding insets show the HAADF images of each nanoprism and the exact positions at which the EEL data were measured. In each case, three main resonances were identified. The energies of these modes vary from one prism to another. Panels (*c*), (*e*) and (*g*) present maps of the intensity distributions of the main resonances detected on the prism in (*a*). Similarly, panels (*d*), (*f*) and (*h*) show the intensity maps of the three modes on the prism in (*b*). For each set of three maps, the common intensity scale is linear and expressed in arbitrary units. The inset above this figure defines the two dimensions *t* and *L*. Reprinted and adapted with permission from (Nelayah *et al* 2010). Copyright 2010 American Chemical Society. The equivalent photon wavelengths were added as insets to the plasmon maps.

sample–vacuum interface (the interface between two media is discussed a little further on). Positioning the beam at the surface or in the bulk of the sample allows for the excitation of distinct modes of oscillation, such as those related to the optical spectrum, including by exciting them with the beam position just outside the surface (so-called 'aloof' spectroscopy) (de Abajo and Howie 1999, Hyun *et al* 2010). Nelayah *et al* (2010) have used this to map the surface plasmons associated with specific colours in silver nanoprisms (figure 3.8).

In the case of interfaces between two media, A and B, the interface plasmon screens the bulk plasmon from A from the bulk plasmon from B, effectively limiting the spatial extent of one bulk plasmon across the interface (the 'begrenzung' effect) (Walls and Howie 1989). This can be used to measure the dielectric properties of atomically thin layers on functional materials, such as WS_2 outer shells on multi-walled carbon nanotubes. It was found that the coating establishes itself as a dielectrically distinct layer when it is more than three monolayers thick (Stolojan *et al* 2005), in agreement with the observation for the SiO_2 gate insulator, which also requires at least three monolayers for its dielectric character to be distinct from the conducting crystalline and amorphous Si either side of it (Muller *et al* 1999).

The bulk plasmon can be used to extract further information about the electronic structure. In the jellium model for electrons in a metal (the electrons are nearly free, with the interaction with the crystal described through an effective mass $m*$), the energy-loss function can be written as (Egerton 2011):

$$\text{Im}\left(\frac{-1}{\varepsilon(E)}\right) = \frac{E\left(\Delta E_p\right)E_p^2}{\left(E^2 - E_p^2\right)^2 + \left(E\Delta E_p\right)^2} \tag{3.4}$$

where the bulk plasmon energy E_p is

$$E_p = \hbar\sqrt{\frac{ne^2}{m*\varepsilon_0}} \tag{3.5}$$

with $\hbar$, e, and ε_0 the reduced Planck's constant, electron charge and vacuum dielectric constant, respectively, and n the density of the valence electrons.

ΔE_p is the width of the plasmon peak in the spectrum, inversely proportional to the plasmon relaxation time. The first implication of (3.4) is that the peak maximum is not the same as the plasmon energy. The peak position is shifted by an amount related to ΔE_p, more pronounced for broader peaks (i.e. shorter relaxation times), such as seen in amorphous carbons. Note that there can also be a further shift due to the size of the aperture used, as the plasmon energy disperses with scattering angle. However, most plasmon spectra contain the very intense ZLP and typically small apertures are used.

Knowledge of the effective mass allows for the calculation of the density of valence electrons from (3.5). For semiconducting and insulating materials, the plasmon energy E_p^i is increased above the free electron value (3.5) by an amount corresponding to the Penn band-gap (an average measure of the bonding-to-antibonding density of states (Penn 1962, Ferrari *et al* 2000a)) E_g:

$$\left(E_p^i\right)^2 \cong E_p^2 + E_g^2. \tag{3.6}$$

For example, the Penn band-gap for c-Si is 1.8 eV (measured 16.7 eV, calculated using (3.5) 16.6 eV (Egerton 2011)). At the same time, diamond's Penn band-gap is 11.9 eV and graphite, with two plasmons at 7 eV and 27 eV, has Penn band-gaps of 11.0 eV and 15.7 eV, respectively.

Nevertheless, the effective mass can be determined as a parameter for certain systems (Ferrari *et al* 2000b), such as carbonaceous materials, and the mass density

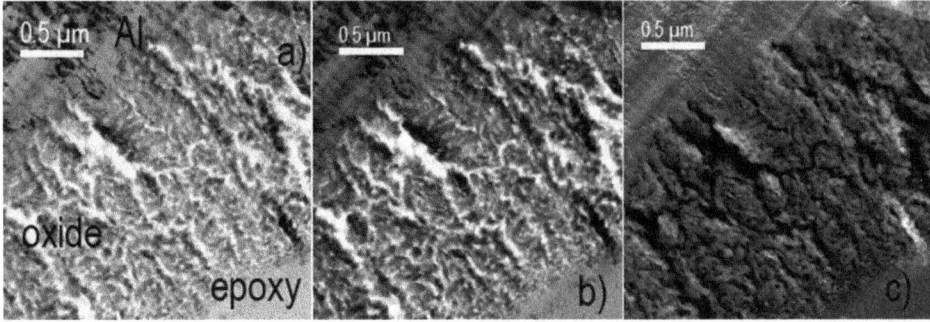

Figure 3.9. A set of (*a*) unfiltered, (*b*) zero-loss images from an Al–epoxy bond, with an Al$_2$O$_3$ bonding layer, prepared by microtoming and (*c*) the resulting relative thickness map, calculated using (3.7). The thickness map shows scratches from the diamond blade in the Al, with dark contrast showing holes in the porous oxide adhesion layers. Bright areas in the oxide layers are not thick, but strong coherence effects that have not been fully removed in the calculation.

of amorphous carbon films can then be related to the sp^3-content, as determined from the carbon K-edge (figure 3.6).

As stated previously, the bulk plasmon peak is the most intense energy-loss peak, indicating that it is also the most likely energy-loss event. We can define therefore λ, the mfp for inelastic scattering, as the average distance between inelastic events as the electron traverses the sample. For Poisson statistics and small scattering angles, we can derive the relative thickness of the sample (accurate to 10% (Zhang *et al* 2012)), the ratio of the sample thickness t, to the mfp λ as:

$$\frac{t}{\lambda} = \ln\left(\frac{I_0}{I_t}\right) \tag{3.7}$$

where I_0 is the integrated intensity of the ZLP and I_t is the integrated intensity of the total spectrum. For Si, the measured mfp is 133 nm (Mayer *et al* 1997), which effectively sets the optimum Si sample thickness for imaging and spectroscopy to be a fraction of this mfp.

If two images are collected, with and without energy filtering, then their log-ratio is effectively a spatial representation of changes in the relative thickness of the specimen (figure 3.9) and can be used to extrapolate thickness information to use in relation to HAADF intensity, in order to be able to interpret the contrast directly as changes in the atomic number. The intensity of the ZLP can be affected by coherence effects, such as strong diffraction by a crystalline grain, and this may also affect the bulk plasmon, as in figure 3.7, therefore the diffraction contrast in both the filtered and the unfiltered images may not be cancelled out when applying (3.7).

3.2.3 EEL spectra processing—the single-scattered distribution

For absolute thickness measurements using the low-loss spectrum in EELS, we can use Kramers–Kronig analysis of the spectrum and its sum rule, which relates the integral of the weighted EEL spectrum to the optical refractive index (at 0 energy loss, which must be known *a priori*) of the sample, with the collection angle and

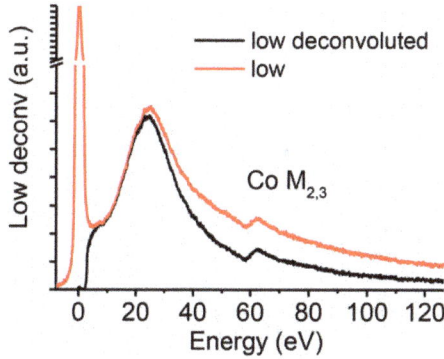

Figure 3.10. A low-loss spectrum and the deconvoluted single-scattering distribution for an amorphous carbon film containing Co nanoparticles. Kramers–Kronig analysis can be performed on the single-scattering distribution (the deconvoluted spectrum) to give the absolute thickness of the film and extract the real and imaginary parts of the dielectric function.

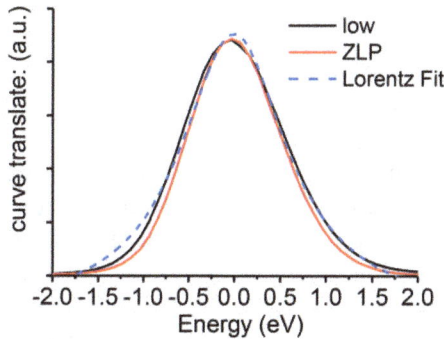

Figure 3.11. The ZLP in a vacuum (red) is narrower than the ZLP in a metallic sample (black). A Lorentz fit to the ZLP in a vacuum (dashed blue line) deviates from the experimental line at the tails and shows that the experimental peak is not symmetric.

incident energy also as input parameters. However, the most critical part of this calculation is that it requires a single-scattering distribution, which is the spectrum corrected for multiple scattering (if the sample thickness is of the order of the mfp for inelastic scattering) and with the ZLP removed. The multiple scattering correction and ZLP removal are performed by deconvoluting the low-loss spectrum and measuring, or modelling, the ZLP (figure 3.10).

The experimental ZLP is obtained by collecting spectra with the beam going through the vacuum, whilst various models can be used to simulate the ZLP (typically Gaussian or Lorentzian functions or combinations thereof, with similar FWHM as the experimental ZLP); these are all implemented in Digital Micrograph™. By far the biggest challenge with this correction is the fact that the ZLP (beam-through-vacuum) is not symmetric, with the level of asymmetry related to the electron tunnelling process leading to emission, among other things (figure 3.11). A monochromator, apart from reducing the chromatic aberration in imaging, also makes the energy profile of the beam symmetric which, along with the very narrow width, makes it

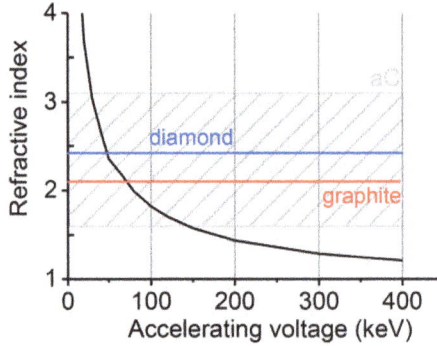

Figure 3.12. Values of the maximum refractive index (black line) above which the velocity of the electron exceeds the speed of light and has to slow down by emitting Čerenkov radiation. Illustrated are the typical ranges for amorphous carbon (hashed region), as well as the values for graphite and diamond. Spectra of graphite above 70 keV will contain Čerenkov radiation.

possible to access optical spectroscopy information with sub-nanometre spatial resolution (Krivanek *et al* 2014, Krivanek *et al* 2013).

Another issue to consider is that the velocity of electrons can, in some cases, come close to the speed of light in a particular medium; for example, 200 keV accelerated electrons traversing any medium with a refractive index higher than ~1.44 have to slow down by emitting Čerenkov radiation (figure 3.12), which can be seen as an increasing background in the region of the optical spectrum and the band-gap. By far the biggest challenge with KKR analysis is the extraction of the ZLP and its tails from the spectrum, a process that is made significantly easier and more accurate by the use of monochromators.

For high-loss edges, such as the C K-edge shown in figure 3.6, there are number of data processing steps that must be undertaken to result in a single-scattered distribution that can then be directly related to (3.3) and (3.4). First there is a pre-edge background, which contains the long tails of all the inelastic loss events that have occurred before the high-loss edge in question and can also be affected by the detector characteristics (non-uniform response) and the microscope conditions (collector aperture size). The visibility of the edge over the background is measured by the ratio of the integrated intensity containing the edge to the intensity of the pre-edge background, integrated over the same energy window. It is called the jump ratio and is used to tune the exposure time to maximise the signal-to-noise ratio. The thicker the sample is, the higher the background and the smaller the jump ratio.

The background can be modelled using a power law, either with a constant exponent (usually between 3 and 4) or with a slowly (linearly) varying exponent (the dashed line in figure 3.13). Generally, the longer the fitting window, the more likely it is that the power law will deviate and introduce errors. The choice of the fitting window for the pre-edge background should, in general, be at least of the same width as the region of interest in the edge; this energy width usually increases as the energy loss increases.

After losing energy to a single-scattering event, the electron will then again experience, with some probability, the same energy losses as appearing in the low-loss spectrum, such as interband transitions and plasmon excitations. This results in a

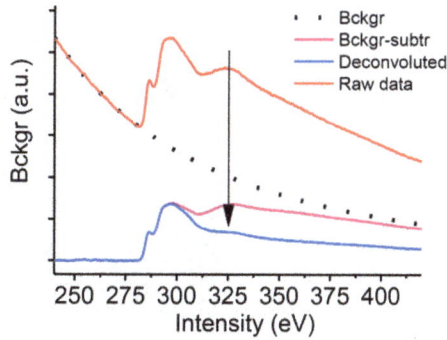

Figure 3.13. A C K-edge from an amorphous carbon sample shown with the data processing steps: power-law background subtraction, followed by the deconvolution of low-energy losses using a pre-measured low-loss spectrum. The arrow points to the bulk plasmon loss appearing after the C K-edge and its successful removal, to result in a single-scattering spectrum distribution.

modulation of the intensity of the edge, with the most visible being the bulk plasmon peak appearing after the edge, indicated by the arrow in figure 3.13. The good news is that these can be removed by deconvolution of the high-loss edge with the low-loss spectrum, after the high-loss edge has been stripped of the pre-edge background. The major sources of error in this process are introduced by the background extrapolation and the fact that low- and high-loss spectra are generally collected with different collection apertures, to compensate for the high dynamic range of the intensity. This generally leads to a ~5% error in the estimates of the composition of samples, although single atoms can be detected in a column of atoms, or in specific cases, of single atom dopants in fullerene molecules (Suenaga *et al* 2000, Colliex *et al* 2012).

3.2.4 High-loss spectra—elemental composition and energy-filtered TEM

The integrated intensity I of the core-loss edge k is proportional to the areal density of an element, N, the incident intensity I and the scattering cross-section for that specific transition, σ_k, and are all functions of the collection angle and the energy integration window:

$$I_k(\beta, \Delta E) \approx NI(\beta, \Delta E)\sigma_k(\beta, \Delta E). \tag{3.8}$$

The scattering cross-sections are evaluated within the analysis software using either a hydrogenic or the Hartree–Slater model, which are relatively accurate for K-edges for integrating energy windows above 20 eV, but less so for other transitions. This means that: (i) ideally edges should be compared over the same energy window ΔE; (ii) the edges should be collected in the same spectrum, rather than splicing two shifted spectra; (iii) ideally the same type of transition (e.g. K-edges) should be compared; and (iv) that the sample is thin of the order of ~0.5λ, the inelastic mpf, giving errors in composition of ~3%–5% (Egerton 2009).

By using an energy-selecting slit in the spectrometer, we can form images with the electrons that have lost a specific energy, allowing us to map the distribution of elements, as discussed in figure 3.4. To make this semi-quantitative, we need to

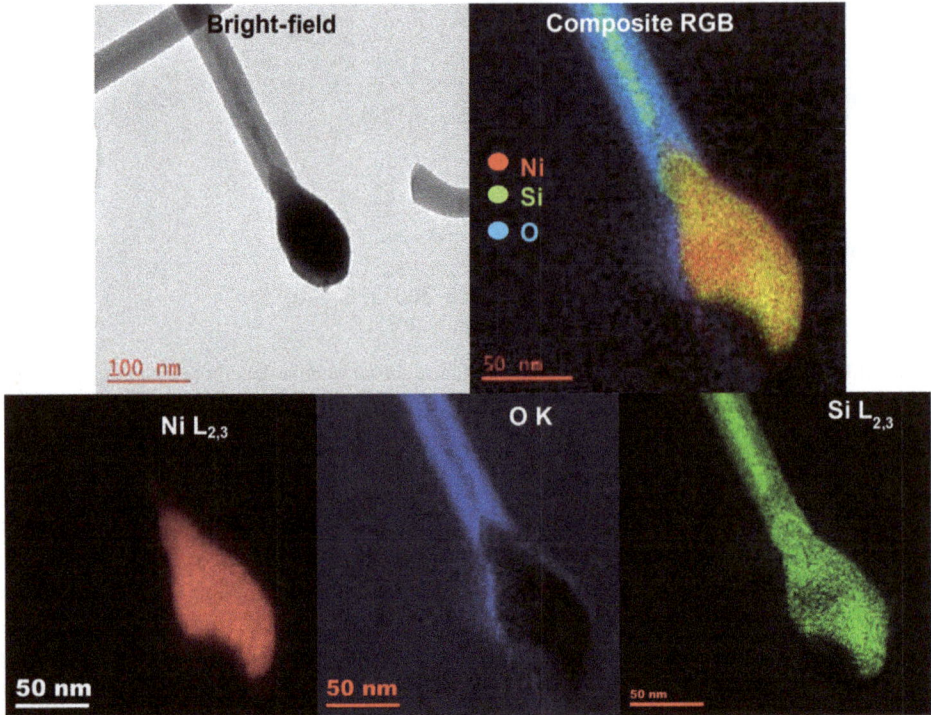

Figure 3.14. A set of bright-field and energy-filtered elemental distribution maps are used to assemble a composite colour map of a Si nanowire grown from a Ni catalyst. It shows the wire is coated with a thick oxide shell, with a Si core, indicating that oxidation has possibly occurred due to oxygen leak during the cool-down of the reactor chamber, as the oxide is too thick for a native oxide (typically 3–5 nm).

estimate the pre-edge background, which is typically achieved by acquiring two pre-edge energy-filtered images and using those to estimate the background. Exposure times can be of the order of minutes and sometimes drift correction is necessary. If the voltage centre of the objective lens of the TEM is not aligned, seen as an apparent shift in the image when changing the high-tension voltage on the TEM, then there will be shifts between images when acquiring EFTEM series, usually accounted for automatically, post-acquisition, by the analysis software. Elemental distribution maps can then be produced, similarly to EDX maps (figure 3.3), as in figure 3.14. The relative composition and distribution of elements can be used to understand the growth processes of nanowires from catalysts, for example showing that the Si diffuses through the catalyst (volume versus surface diffusion) and that there is a thick oxide sheath, that can either be used as a gate insulator in field-emission transistor applications, or it can become an impediment if trying to contact the Si wire itself. It is worth noting that microscopy examines these nanomaterials after the reaction has finished and sometimes the processes occurring during the termination of the reaction and the cool down of the catalyst, end up dominating the final product. For example, Ni and Si can phase-separate upon cooling down, leading to a possible conclusion that Si diffuses on the outside of the catalyst. This is why *in situ* TEM is an essential tool for elucidating important processes such as carbon nanotube growth (Helveg *et al* 2004, Baker *et al* 1972) and Si nanowire growth (Baker and Thomas 1972, Ross 2010).

As with EDX, quantitative mapping requires the acquisition of point-by-point spectra (known as spectrum imaging), followed by the background subtraction and deconvolution routines described before. This acquisition requires automatic spatial and energy drift correction and data processing, as well as collecting low-loss spectra at the same positions, so the demands on sample stability are significant. Nevertheless, with aberration correction and very thin samples, atomic-scale compositional mapping is possible in short times, due to the increased intensity in the beam (Muller *et al* 2008), as seen in figure 1.5.

Upon exiting the scattering atom, the electron propagates as a wave, which means it can scatter from the neighbouring atomic shells, resulting in coherence effects. These manifest as an extended energy-loss fine structure (EXELFS), long range oscillations in the EEL spectrum that redistribute the intensity in the edge and are particularly pronounced in oxides. In figure 3.15, the experimental spectrum is shown to drop below the level of the Hartree–Slater theoretical continuum, particularly at ~730 eV. These oscillations make it difficult to extract the pre-edge background when two edges are close together, such as the Ti $L_{2,3}$-edge (454 eV) and the O K-edge (532 eV) (also Cr and Fe oxides), making quantification of titania compositions difficult. Instead, we look at the fine structure in the edges themselves (degeneration, relative heights and integrated intensities of the peaks and energy shifts) to fingerprint the particular compound (Brydson *et al* 1993). As the intensity in the edge is redistributed to higher energies, it also means that the choice of the integration window for extracting relative compositions will influence the results significantly.

On the other hand EXELFS can be used, just like its synchrotron counterpart, extended x-ray absorption fine structure, to understand the local atomic structure and extract the structure factor around specific atoms, as selected by the high-loss edge (Batson 2008, Sikora *et al* 2000).

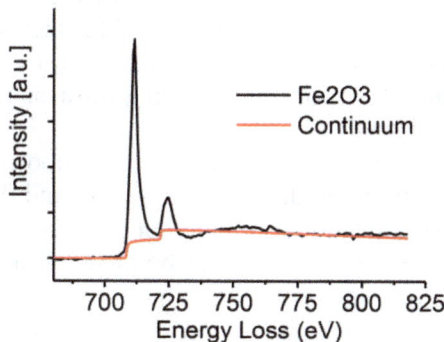

Figure 3.15. The intensity in the $L_{2,3}$ white line is redistributed by coherent scattering of the scattered electron into broad oscillations about the continuum (red line), which makes quantification sensitive to the width of the energy window selected for integration. This also has implications on the measurement of the occupancy of the 3d band.

3.2.5 The energy-loss near edge structure (ELNES)

The near-edge structure of the high-loss edges contains information about the atomic bonding, the number of unoccupied states in the final state and is generally localised to the size of the final state orbital; thus, it is possible to perform atomic-resolution compositional and bonding mapping by EELS (Muller *et al* 2008). The one challenge still remaining for EELS is to measure the onset of the transition edge with the accuracies available with x-ray photoelectron spectroscopy (XPS) and x-ray absorption spectroscopy. This is because chemical bonding, or changes in chemical bonding, usually result in a shift in the (initial) core level, which means that the transition energy changes depending on the chemistry. The engineering difficulty is that we measure the energy loss with respect to the acceleration energy, which is of the order of 10^5 eV; this means we need stabilities of the order of 10–100 ppb in voltage supplies of the microscope gun, the lenses and the spectrometer to measure with meV accuracies, particularly when the exposure times reach several tens of seconds. However, there have been significant advances in chromatic aberration correction which have made 10 meV achievable (Krivanek *et al* 2014), based on a monochromator with current-beam sensing at the energy selecting aperture (Krivanek *et al* 2009).

In the absence of such a monochromator, the only alternative is to calibrate the peak energies in relation to known peaks, such as the amorphous carbon π^* peak at 285.5 eV (figure 3.6), as is also the practice in XPS. This means that the peak of interest and the C K-edge of amorphous carbon need to be acquired at the same time, within the same spectrum. The other point to note is that the peak position can be determined with better accuracy than the spectrum x-axis channel energy width by using several data points to fit the peak, with the peak position error scaling down with the square-root of the number of points used for the fitting (assuming Poisson statistics and a good fit[5]). This means that, for edges acquired simultaneously (e.g. ELSP), we can measure core-level shifts down to a few meV.

Other approaches have used a beam current sensing approach to locate and fix, in a feed-back loop, the position of the ZLP within the spectrometer (Bleloch *et al* 1999) and designing a dual detector that switches rapidly between the low-loss spectrum and the high-loss spectrum, with the two detectors set up for the different dynamic ranges of the two types of spectra. The ZLP then becomes the reference energy and this approach also solves the issue of recording a low-loss spectrum at the same position as the high-loss spectrum, allowing for accurate deconvolution and quantification (Scott *et al* 2008).

The area under peaks in the near-edge structure is proportional to the number of states of that particular symmetry, the density of atoms and the matrix element that defines the probability of transition as a function of orientation (e.g. in the case of an anisotropic material) (see (3.3)). For the carbon K-edge (figure 3.6), we can fit the π^* peak (transitions from 1s to the 2sp-hybridised π-antibonding states) to extract

---

[5] A 'good fit' here is that the function is a good physical approximation of the peak, particularly when the peak is asymmetric (see figure 3.11). If the fit is not 'good' then there is an option to reduce the fitting energy window (reducing the accuracy) until the fit becomes 'good'; this is an experimental choice.

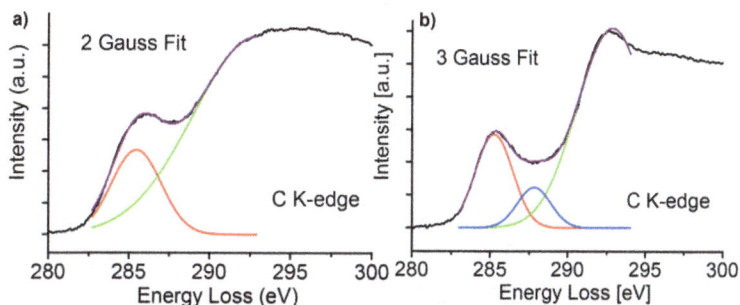

Figure 3.16. (*a*). The π^* peak in the C K-edge of an amorphous carbon can be separated from the σ^* peak by modelling it with two Gaussians. The area under the red curve, normalised to a specific total area of the C K-edge and normalised to the equivalent 100% sp^2-bonded quantity gives the relative sp^2-content of the sample. (*b*) For more graphitic samples, with a pronounced ledge between the π^* and the σ^* peaks, as well as a more defined σ^* peak, a third Gaussian (blue line) is required to model the area of the π^* peak.

the concentration of sp^2-bonded carbon in an amorphous carbon sample, assuming that it is only composed of sp^2- and sp^3-type bonds. In some cases, we can extract the π^* peak area by fitting with just two Gaussians, see figure 3.16, and normalising the π^* peak area to the area of the C K-edge in the range [283–294] eV; this normalised area in a 100% sp^2-bonded sample (graphite, C_{60}) is 0.309, so comparing the two normalised areas returns the relative sp^2-content (Ferrari *et al* 2000b).

In some cases, however, particularly more graphitic samples, a third Gaussian centred on 288.1 eV, (see figure 3.16), is required to fit the π^* peak contribution, and the equivalent normalised area is 0.33 (Papworth *et al* 2000).

In 3d-transition metals, the $L_{2,3}$-peaks[6] show two sharp peaks on top of a continuum type local density of states (figure 3.15), which relate to the 3d–4s hybridisation, where the 3d electrons are strongly localised and the 4s electrons are delocalised. The area of the $L_{2,3}$-peaks is proportional to the occupancy of the 3d orbital (Pearson *et al* 1988, Pearson *et al* 1993), so the area of the peaks can be calibrated and used to determine the oxidation state of the sample (the higher the valence, the more the empty states in the 3d band, the higher the area) (figure 3.17). However, we must not forget also that oxides can experience damage quickly due to electron-beam irradiation, so a time-resolved series of spectra of the white lines can and should be used to assess damage to the sample. This damage is not always immediately visible with imaging techniques (e.g. in titania or alumina) but is readily visible in the O K-edge spectra as a change in oxidation co-ordination and state.

It can also be used to monitor small changes in the occupancy brought on by impurities at grain boundaries, such as C and P in Fe, where small exchanges of electrons between the localised 3d and delocalised 4s characters can lead to a change in the number of bonding and antibonding occupied energy levels (as well as meV shifts in the peak energies). This can then relate to the cohesion energies (Muller *et al* 1995, Stolojan *et al* 2001) and the type of failure of the sample (brittle or ductile), for example.

---

[6] The $L_{2,3}$-edge is also commonly referred to as 'white lines' due to their appearance on old photographic recording media of x-ray and EEL spectra.

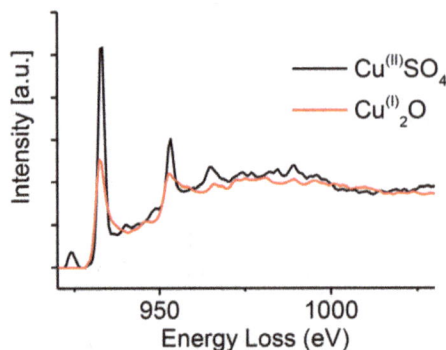

Figure 3.17. Cu does not show white lines, since 3d is fully occupied, but its oxides (Cu(I) and Cu(II)) show white lines, with their integrated intensities proportional to the oxidation state. The spectra are aligned to the same onset energy.

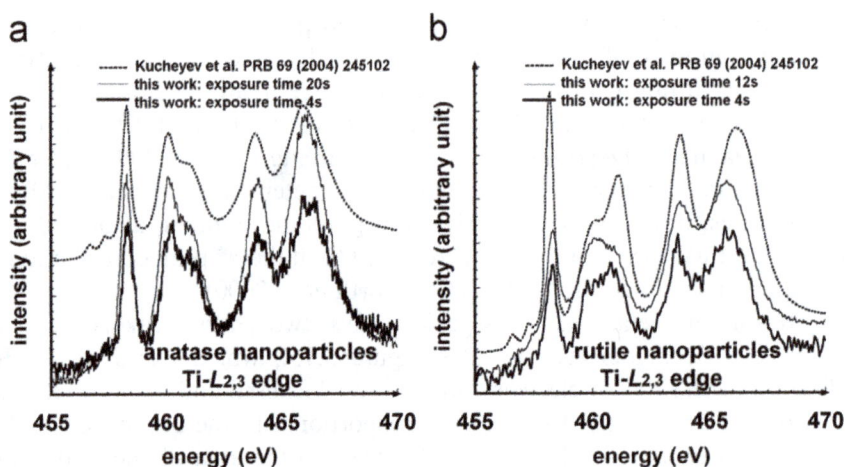

Figure 3.18. A monochromated, aberration-corrected instrument can differentiate between the (*a*) anatase and (*b*) rutile phases of titania by looking at the relative ratios of the peaks around 460 eV. Reproduced with permission from (Cheynet *et al* 2010). Copyright 2010 Elsevier.

The ratio of the L_3- to L_2-peaks[7] should reflect the 2:1 ratio of the population of the spin–orbit split 2p level, however, it can vary from as low as 0.7 for Ti to 3:1 for Fe and Ni due to many-electron effects, such as magnetism (Leapman and Grunes 1980).

The co-ordination of the atom and its oxidation state can also be inferred by recognising degeneracies in peaks, combined with *ab intio* calculations (Brydson *et al* 1993, DeGroot and Kotani 2008, Tan *et al* 2012). With aberration correction and a monochromator, the relative ratios of the peaks can be used to identify, for example, two co-ordination variants of the same oxidation state, namely the rutile and anatase phases for TiO_2 (Cheynet *et al* 2010).

---

[7] Also known as the 'white line ratio', or the 'branching ratio'.

As stated earlier in this chapter, the future goal for EELS is to be able to measure core-level shifts with sub-nanometre resolution, where the cases presented in figures 3.17 and 3.18 would be helped by knowing the onset energies. For example, a 400 meV shift is expected to be observed between anatase and rutile in the Ti $L_{2,3}$-edge (Scanlon *et al* 2013). The advances made by Krivanek *et al* (2014) are ground-breaking and it is only a short matter of time before the incredible energy resolution is complemented by similar improvements in energy stability over the longer exposure times required for good statistics in the ELNES spectra.

References

Baker R T K, Barber M A, Waite R J, Harris P S and Feates F S 1972 Nucleation and growth of carbon deposits from nickel catalyzed decomposition of acetylene *J. Catal.* **26** 51

Baker R T K and Thomas R B 1972 Continuous microscopic observation of reaction of silicon with methane in presence of iron *J. Cryst. Growth* **12** 185

Batson P E 2008 Local crystal anisotropy obtained in the small probe geometry *Micron* **39** 648–52

Berger S D, McKenzie D R and Martin P J 1988 EELS analysis of vacuum arc-deposited diamond-like films *Phil. Mag. Lett.* **57** 285–90

Bethe H 1929 Over the passage of cathode rays by lattice electric fields *Z. Phys.* **54** 703–10

Bleloch A, Brown L M, Marsh M J, McMullan D, Rickard J J and Stolojan V 1999 Cancelling energy drift in a PEELS spectrometer *Electron Microsc. Anal.* **1999** 195–8

Bright D S and Newbury D E 1991 Concentration histogram imaging—a scatter diagram technique for viewing 2 or 3 related images *Anal. Chem.* **63** A243

Brown L M 1997 A synchrotron in a microscope *Biennial Mtg of the Electron Microscopy and Analysis Group of the Institute of Physics (2–5 September 1997, Cambridge, UK)* (Bristol: IOP) pp 17–22

Brydson R, Garvie L A J, Craven A J, Sauer H, Hofer F and Cressey G 1993 $L_{2,3}$ edges of tetrahedrally coordinated d^0 transition-metal oxyanions XO_4^{n-} *J. Phys.: Condens. Matter* **5** 9379–92

Cheynet M, Pokrant S, Irsen S and Kruger P 2010 New fine structures resolved at the ELNES Ti-L-2,L-3 edge spectra of anatase and rutile: comparison between experiment and calculation *Ultramicroscopy* **110** 1046–53

Colliex C, Gloter A, March K, Mory C, Stephan O, Suenaga K and Tence M 2012 Capturing the signature of single atoms with the tiny probe of a STEM *Ultramicroscopy* **123** 80–9

Daniels H, Brown A, Scott A, Nichells T, Rand B and Brydson R 2003 Experimental and theoretical evidence for the magic angle in transmission electron energy loss spectroscopy *Ultramicroscopy* **96** 523–34

de Abajo F J G and Howie A 1999 Electron spectroscopy from outside—aloof beam or near field? *Biennial Mtg of the Electron Microscopy and Analysis Group of the Institute of Physics (24–27 August 1999, Sheffield, UK)* (Bristol: IOP) pp 327–30

DeGroot F and Kotani A 2008 Core level spectroscopy of solids *Core Level Spectrosc. Solids* **6** 1–490

Egerton R 2011 *Electron Energy-Loss Spectroscopy in the Electron Microscope* (New York: Springer)

Egerton R F 2009 Electron energy-loss spectroscopy in the TEM *Rep. Prog. Phys.* **72** 016502

Eibl O 1993 New method for absorption correction in high-accuracy, quantitative EDX micro-analysis in the tem including low-energy x-ray-lines *Ultramicroscopy* **50** 179–88

Ferrari A C, Kleinsorge B, Adamopoulos G, Robertson J, Milne W I, Stolojan V, Brown L M, LiBassi A and Tanner B K 2000a Determination of bonding in amorphous carbons by electron energy loss spectroscopy, Raman scattering and x-ray reflectivity *J. Non-Cryst. Solids* **266** 765–8

Ferrari A C, Libassi A, Tanner B K, Stolojan V, Yuan J, Brown L M, Rodil S E, Kleinsorge B and Robertson J 2000b Density, sp^3 fraction, and cross-sectional structure of amorphous carbon films determined by x-ray reflectivity and electron energy-loss spectroscopy *Phys. Rev.* B **62** 11089–103

Gloter A, Chu M W, Kociak M, Chen C H and Colliex C 2009 Probing non-dipole allowed excitations in highly correlated materials with nanoscale resolution *Ultramicroscopy* **109** 1333–7

Helveg S, Lopez-Cartes C, Sehested J, Hansen P L, Clausen B S, Rostrup-Nielsen J R, Abild-Pedersen F and Norskov J K 2004 Atomic-scale imaging of carbon nanofibre growth *Nature* **427** 426–9

Hyun J K, Levendorf M P, Blood-Forsythe M, Park J and Muller D A 2010 Relativistic electron energy loss spectroscopy of solid and core–shell nanowires *Phys. Rev.* B **81** 165403

Jouffrey B, Schattschneider P and Hebert C 2004 The magic angle: a solved mystery *Ultramicroscopy* **102** 61–6

Kirkland E J 2005 Some effects of electron channeling on electron energy loss spectroscopy *Ultramicroscopy* **102** 199–207

Krivanek O L *et al* 2014 Vibrational spectroscopy in the electron microscope *Nature* **514** 209

Krivanek O L, Lovejoy T C, Dellby N and Carpenter R W 2013 Monochromated STEM with a 30 meV-wide, atom-sized electron probe *Microscopy* **62** 3–21

Krivanek O L, Ursin J P, Bacon N J, Corbin G J, Dellby N, Hrncirik P, Murfitt M F, Own C S and Szilagyi Z S 2009 High-energy-resolution monochromator for aberration-corrected scanning transmission electron microscopy/electron energy-loss spectroscopy *Phil. Trans. R. Soc.* A **367** 3683–97

Leapman R D and Grunes L A 1980 Anomalous L_3–L_2 white-line ratios in the 3D transition-metals *Phys. Rev. Lett.* **45** 397–401

Lewis E A, Slater T J A, Prestat E, Macedo A, O'Brien P, Camargo P H C and Haigh S J 2014 Real-time imaging and elemental mapping of AgAu nanoparticle transformations *Nanoscale* **6** 13598–605

Litvinenko K L *et al* 2015 Coherent creation and destruction of orbital wavepackets in Si:P with electrical and optical read-out *Nat. Commun.* **6** 8

Mayer J, Eigenthaler U, Plitzko J M and Dettenwanger F 1997 Quantitative analysis of electron spectroscopic imaging series *Micron* **28** 361–70

Menon N K and Yuan J 1998 Quantitative analysis of the effect of probe convergence on electron energy loss spectra of anisotropic materials *Ultramicroscopy* **74** 83–94

Millar L, Taherparvar H, Filkin N, Slater P and Yeomans J 2008 Interaction of $(La_{1-x}Sr_x)_{(1-y)}MnO_3$–$Zr_{1-z}Y_zO_{2-d}$ cathodes and $LaNi_{0.6}Fe_{0.4}O_3$ current collecting layers for solid oxide fuel cell application *Solid State Ion.* **179** 732–9

Muller D A, Kourkoutis L F, Murfitt M, Song J H, Hwang H Y, Silcox J, Dellby N and Krivanek O L 2008 Atomic-scale chemical imaging of composition and bonding by aberration-corrected microscopy *Science* **319** 1073–6

Muller D A, Sorsch T, Moccio S, Baumann F H, Evans-Lutterodt K and Timp G 1999 The electronic structure at the atomic scale of ultrathin gate oxides *Nature* **399** 758–61

Muller D A, Subramanian S, Batson P E, Sass S L and Silcox J 1995 Near atomic scale studies of electronic structure at grain boundaries in Ni_3Al *Phys. Rev. Lett.* **75** 4744–7

Nelayah J, Kociak M, Stephan O, Geuquet N, Henrard L, de Abajo F J G, Pastoriza-Santos I, Liz-Marzan L M and Colliex C 2010 Two-dimensional quasistatic stationary short range surface plasmons in flat nanoprisms *Nano Lett.* **10** 902–7

Papworth A J, Kiely C J, Burden A P, Silva S R P and Amaratunga G A J 2000 Electron-energy-loss spectroscopy characterization of the sp^2 bonding fraction within carbon thin films *Phys. Rev.* B **62** 12628–31

Pearson D H, Ahn C C and Fultz B 1993 White lines and d-electron occupancies for the 3d and 4d transition-metals *Phys. Rev.* B **47** 8471–8

Pearson D H, Fultz B and Ahn C C 1988 Measurements of 3d state occupancy in transition-metals using electron-energy loss spectrometry *Appl. Phys. Lett.* **53** 1405–7

Penn D R 1962 Wave-number-dependent dielectric function of semiconductors *Phys. Rev.* **128** 2093

Ritchie R H 1957 Plasma losses by fast electrons in thin films *Phys. Rev.* **106** 874–81

Ross F M 2010 Controlling nanowire structures through real time growth studies *Rep. Prog. Phys.* **73** 114501

Saito M, Kimoto K, Nagai T, Fukushima S, Akahoshi D, Kuwahara H, Matsui Y and Ishizuka K 2009 Local crystal structure analysis with 10-pm accuracy using scanning transmission electron microscopy *J. Electron Microsc.* **58** 131–6

Scanlon D O *et al* 2013 Band alignment of rutile and anatase TiO_2 *Nat. Mater.* **12** 798–801

Scott J, Thomas P J, MacKenzie M, McFadzean S, Wilbrink J, Craven A J and Nicholson W A P 2008 Near-simultaneous dual energy range EELS spectrum imaging *Ultramicroscopy* **108** 1586–94

Sikora T, Hug G, Jaouen M and Rehr J J 2000 Multiple-scattering EXAFS and EXELFS of titanium aluminum alloys *Phys. Rev.* B **62** 1723–32

Spence J C H and Tafto J 1983 ALCHEMI: a new technique for locating atoms in small crystals *J. Microsc.* **130** 147–54

Stolojan V, Brown L M and McMullan D 2001 High energy resolution measurements of the density of states at ferrous grain boundaries by EELS *Electron Microscopy and Analysis* ed M Aindow and C J Kiely (Bristol: IOP Publishing)

Stolojan V, Silva S R P, Goringe M J, Whitby R L D, Hsu W K, Walton D R M and Kroto H W 2005 Dielectric properties of WS_2-coated multiwalled carbon nanotubes studied by energy-loss spectroscopic profiling *Appl. Phys. Lett.* **86** 063112

Suenaga K, Tence T, Mory C, Colliex C, Kato H, Okazaki T, Shinohara H, Hirahara K, Bandow S and Iijima S 2000 Element-selective single atom imaging *Science* **290** 2280

Tan H, Verbeeck J, Abakumov A and Van Tendeloo G 2012 Oxidation state and chemical shift investigation in transition metal oxides by EELS *Ultramicroscopy* **116** 24–33

Tiemeijer P C 1999 Operation modes of a TEM monochromator *Biennial Mtg of the Electron Microscopy and Analysis Group of the Institute of Physics (24–27 August 1999, Sheffield, UK)* pp 191–4

Walls M G and Howie A 1989 Dielectric theory of localized valence energy-loss spectroscopy *Ultramicroscopy* **28** 40–2

Walther T 2003 Electron energy-loss spectroscopic profiling of thin film structures: 0.39 nm line resolution and 0.04 eV precision measurement of near-edge structure shifts at interfaces *Ultramicroscopy* **96** 401–11

Watanabe M and Williams D B 2006 The quantitative analysis of thin specimens: a review of progress from the Cliff–Lorimer to the new zeta-factor methods *J. Microsc.* **221** 89–109

Williams D B and Carter C B 2009 *Transmission Electron Microscopy: A Textbook for Materials Science* (New York: Springer)

Xin H L L, Dwyer C and Muller D A 2014 Is there a Stobbs factor in atomic-resolution STEM-EELS mapping? *Ultramicroscopy* **139** 38–46

Zhang H R, Egerton R F and Malac M 2012 Local thickness measurement through scattering contrast and electron energy-loss spectroscopy *Micron* **43** 8–15

www.ingramcontent.com/pod-product-compliance
Lightning Source LLC
Chambersburg PA
CBHW081554220326
41598CB00036B/6672